PRACTICAL LABORATORY PLANNING

Reproduced by courtesy of the Ian Clunies Ross Memorial Foundation

This mural is in the entrance foyer of Clunies Ross House, National Science Centre, Melbourne. It was presented by the Australian Veterinary Association as a memorial to one of its most distinguished members, Sir Ian Clunies Ross, who held many senior research and academic appointments and was Chairman of CSIRO from 1949 until his death in 1959. The three panels of the mural depict Clunies Ross in the roles of scientist, administrator, and communicator — an educated, civilized man confronting ignorance symbolized by the dark, untamed, potentially dangerous beast.

The mural covers an area of approximately 100 square feet and is painted on $\frac{3}{4}$-inch thick waterproof coachwood plywood. The artist, Robert Ingpen, used specially selected inorganic paint, and the final surface was lightly polished with silicone wax.

PRACTICAL LABORATORY PLANNING

W. R. FERGUSON
B.E. (Adel.), F.S.A.S.M. (Arch.)
Laboratory Design Consultant (formerly with Commonwealth Scientific and Industrial Research Organization, Australia)

A HALSTED PRESS BOOK

JOHN WILEY & SONS
New York—Toronto

PUBLISHED IN THE U.S.A. AND CANADA BY
HALSTED PRESS
A DIVISION OF JOHN WILEY & SONS, INC., NEW YORK

Library of Congress Cataloging in Publication Data

Ferguson, William Rex
Practical laboratory planning

"A Halsted Press book"
1. Laboratories—Design and construction. I. Title.

TH4652.F47 727'.5 72–10256
ISBN 0–470–25750–4

WITH 61 ILLUSTRATIONS AND 3 TABLES

Printed in Great Britain by Galliard Limited, Great Yarmouth, Norfolk, England.

Foreword

In a research organisation the functional design of laboratories is of basic importance. It affects efficiency in numerous ways, is reflected in the operational costs, places limits on flexibility and is a very important element in the development of safe working practices.

By the mid-1960's CSIRO had accumulated considerable information and experience relating to laboratory design and it was thought that the time was opportune to assemble this in a form which would be readily available to people within and outside the Organization. Accordingly, the Executive asked the author to undertake the preparation of a publication on the design of laboratory buildings, incorporating all CSIRO's experience in this field.

Publication of this book represents the completion of that assignment.

Rex Ferguson must be one of the few people who have spent almost a lifetime designing and building research laboratories. He graduated in architectural engineering at the University of Adelaide in 1931. Work for architects was very scarce in those depression years, so for him there followed an enforced but valuable period working for a building contractor as a cost clerk and supervisor.

After studying for two years in the USA and England on a travelling studentship, he joined CSIRO in 1935 as an assistant research officer to work on timber utilisation problems in the Division of Forest Products. A few years later he moved to the Organization's Head Office to become CSIRO Architect, and he continued in this area of work until his retirement in 1970.

His international reputation in the field of laboratory design has led to invitations to act as consultant on projects in Bangkok, Colombo, Djakarta, Honolulu and Montevideo.

This book has been written to assist all those involved in creating a finished laboratory—the scientist, the project officer, the architect, the engineer and the builder. Rex Ferguson is extraordinarily well qualified to advise such people on practical laboratory planning and, in writing this book, he has made his and CSIRO's experience available to all.

J. R. PRICE
Chairman, CSIRO

Introduction

The explosive growth in science and technology during the past fifty years has generated increasing demands for research; there has been a tremendous increase in both the number and size of research laboratories, and at present there certainly is no indication that there will be any reduction in this rate of expansion. As the economies of most countries become increasingly geared to technological advance, the demand for even larger research facilities must grow. The stern warning issued in 1966 by Dr James R. Killian Jr., Chairman of the Massachusetts Institute of Technology, is still true today:

> 'First is the danger, given our present strength, that we may rest on our oars, thinking that the race is won. Actually, we may be only at the beginning of unexampled scientific and engineering achievement, on the threshold of an Augustan age of unparalleled creativity. With the future so promising, this is not the time to relax our scientific effort or for timid talk about having reached some kind of ceiling in our upsurge of scientific and technological strength.'

As the needs of the scientist become more exacting, laboratory planning is developing into a complex and specialised field. Despite the greater attention being given to the design and construction of laboratories, the unfortunate truth is that mistakes seem always to occur. In an effort to eliminate these defects in future buildings, both client and architect should become more conversant with the subject.

This book is intended as a guide for people who are planning chemistry and physics research laboratories; routine chemistry and student laboratories are not dealt with specifically, but the same principles of layout and services can usually be applied. The book deals with the importance of effective communication between client and architect, the value of preliminary planning and the role of the project officer; it discusses the size and layout of individual laboratories, the design of laboratory fittings and services, fire and safety precautions and the many things to be considered in the design and layout of the building.

Descriptions of a number of CSIRO and overseas laboratories have been included, together with technical details of many features; these are given in the hope that some, at least, will be of particular interest to the reader. The buildings vary greatly in size, but all are examples of successful and practical laboratory planning.

Finally a note on units. In most cases British measurements are quoted, although it has not been possible to be entirely consistent, particularly when dealing with laboratories designed to a metric module. Nevertheless, all costs are quoted in dollars Australian ($A) so suitable adjustments will need to be made to the figures for US and European laboratories.

Contents

Acknowledgements

There are many people I wish to thank:

The Chairman and members of the Executive of the Commonwealth Scientific and Industrial Research Organization for giving me the opportunity to write this book, both before and after my retirement; Mr B. Beresford Smith, Assistant Secretary (Works and Buildings), for his support and encouragement; a number of scientists whose opinions I always have sought and value highly; other members of CSIRO for advice and for assistance with the illustrations. I also wish to express my appreciation to the Directors of the overseas laboratories which are mentioned; they were all extremely co-operative in supplying information and granting me permission to use it. Finally, I am most grateful to my secretary, Margaret Allan; without her competence and diligence, I doubt if this book would ever have seen the light of day.

W. R. Ferguson

Conversion Table

SI Equivalents

1 in = 25·4 mm
1 ft = 305 mm
1 ft^2 = 0·093 m^2
1 acre = 0·405 ha
1 lb = 0·454 kg
1 lb/in^2 = 703 kg/m^2
1 lb/ft^2 = 4·88 kg/m^2
1 cwt = 50·8 kg
1 long ton = 1·016 metric tons

1 ft^3 = 0·028 m^3
1 gal = 4·55 litres
1 ft/min = 0·005 m/s
1 mph = 1·609 km/h
1 hp = 746 W
1 lm/ft^2 = 10·76 lx
1 Btu/h – 0·293 W
1 $Btuh/ft^2$ = 41 MW/m^2

Chapter 1

Responsibility of the Client

Few people appreciate the significance of the client's role in the design and construction of a building—sometimes not even the client himself! No matter how competent the architect, there still is much that the client can contribute, especially in a large and complex project such as a laboratory; it is therefore in his own interests to realise this and accept the responsibility involved. The situation calls for a maximum team effort by the management, scientist, project officer, architect, engineer and builder.

The selection of the architect is one of the first major decisions. I think it is wise to choose one who has had previous experience in the specialised field of laboratory design; furthermore, a small firm is likely to be more co-operative and flexible in their approach than a large firm. An alternative is to engage a design–construction organisation. In recent years these people have challenged the traditional role of the architect; they offer an all-inclusive service covering the architectural and engineering design, and the construction of the building—all in the one contract. It is assumed that, in the process of selection, the client will make enquiries regarding several architects and visit some of their completed buildings, in order to satisfy himself as to their style and the standard of their work.

The client's preplanning and liaison with the architect is extremely important; the more thoroughly he does this, the better are his prospects of a completely satisfactory building. Just how this is arranged depends on the size of the current project and the extent of probable future buildings. Most large organisations employ their own group of architects and engineers who have become skilled in laboratory design; smaller organisations appoint a senior member of staff as project officer, or engage a laboratory design consultant. In this way, a strong and efficient link between client and architect is provided.

It would be foolish to give the impression that liaison between client and architect always operates smoothly; with the multitude of problems which inevitably arise during the construction of any building, and a possible clash of personalities, it is not surprising that friction can develop. Some architects resent suggestions made by the client because they regard them as interference; but it must be acknowledged that the client can have valid reasons for insisting that some features be incorporated, or others omitted. Liaison is indeed a situation calling for the utmost goodwill and co-operation on both sides.

BUILDING SECTION

Preplanning and liaison can best be undertaken by a specialist group within the client's own organisation. This Building Section prepares preliminary plans, detailed laboratory layouts including fittings, the building brief and is also responsible for liaison between the organisation and the designing architect. Rather than engage an outside architect, the management sometimes prefers to set up this section so that it can handle the actual design and supervision of their buildings—including the laboratory fittings and services. Another important function is to advise the management on technical matters such as selection and development of sites, estimates of costs and programming.

The staff of this section generally are professionally qualified in architecture or engineering, and should have extensive knowledge and practical experience of laboratory design. An architecturally trained project officer has the advantage of being more conversant with the problems of architectural design and planning, and of building construction and costs. However, mechanical and electrical services are figuring so prominently in laboratories nowadays, that it is most desirable for a qualified engineer also to be on the staff to advise the project officer on these matters.

PROJECT OFFICER

As soon as a new building is first mooted, an officer of the Building Section should be assigned the duties of project officer. If there is no Building Section, then a member of the staff should be appointed to act in this capacity; he may be an executive, a laboratory worker or any person with sufficient enthusiasm and commonsense. It is as well to remember that the work of a project officer might start as a part-time extra duty, but it can quickly become a full-time activity.

Whilst it is necessary for the project officer to discuss detailed requirements of individual laboratories with the scientists concerned, it is best if information on the broader issues is channelled to him through one particular scientist to whom this duty is delegated; for a large project, it might be preferable to have a small committee. In order to ensure that everyone has an opportunity to offer comment or criticism, it is desirable to have occasional discussions with the entire staff.

Most scientists readily consider suggestions and will conform to a certain amount of standardisation; unfortunately there are others who, without justification, insist that their individual requirements be met, even if these are excessive or unnecessarily costly. It is important, therefore, that the project officer has not only a sympathetic approach to the needs of the scientist, but also the ability to persuade him to accept alternatives in instances where these will be more practicable and, in many cases, cheaper.

The project officer must spend many hours in discussions, attend site meetings and make frequent and regular inspections during construction; it is, of course, very important to keep detailed notes of all these meetings. He is expected to maintain time schedules both on the drawing board and on the site; in fact, he must be capable of exerting a constant gentle pressure on all parties involved. The secret of success is to be like a duck—smooth and unruffled on top, but paddling like hell underneath.

PREPLANNING

Those who already have experience in the design and construction of laboratories are well aware of the tremendous amount of preplanning which should be done before the architect commences.

Ideally it should be started as soon as the project is first suggested; even if some considerable time elapses before the job proceeds, it is nevertheless a great advantage to have layout drawings available as a basis for discussion. There is no doubt that more thought over a longer period will result in a better laboratory. It is a great mistake to let the months slip by without any preplanning because, almost overnight, a situation can develop where the building is urgently required or finance is allocated sooner than anticipated; if this happens, then the architect is expected to meet a tight schedule and he has to proceed without adequate information. There is no substitute for preplanning in detail.

It is essential that the project officer be conversant with current developments in laboratory design; but this is not easy because nowadays so many different types of laboratories are being built, and new materials and techniques are becoming available. The most effective method of keeping up to date is to visit other laboratories. A new idea or a different slant on a problem can quite often be obtained by talking with the laboratory manager or several of the scientists, many of whom take a lively interest in laboratory design. They are very willing to discuss the good features, and are generally quick to point out the bad ones; sometimes, however, they are reluctant to mention mistakes and, in these circumstances, several well directed questions usually evoke valuable information. In the case of the National Bureau of Standards Laboratories described on p. 129 a seven-man Laboratory Planning Committee visited 11 of the best laboratories in the country on a fact-finding mission. This might appear to have been an extravagance, but it was a huge project and I am sure that the visits showed a handsome dividend. Even for quite small projects such as a single isotope laboratory, it is important to check on current trends.

Individual impressions of a laboratory or some feature of it, expressed in terms of good or bad, are of little help to the project officer; specific details are needed, and they should be systematically recorded. It is suggested, therefore, that a questionnaire be used. This should include the following points:

Name of laboratory:
Location:
Type of research:
Date completed:
Number of buildings in research group; if more than one, why?:
Orientation:
Type and size of window per module:
Number of storeys:
Floor-to-ceiling height:
Floor-to-floor height:
Laboratory module:
Laboratory depth:
Bench height:
Bench width:
Bench top material and finish:
Extent of standardisation of laboratory layouts and fittings:
Number of occupants per standard laboratory:
Size and number of laboratory offices and location in relation to laboratories:
What services are provided and in what position are they reticulated?:

Number of outlets per bench:
Is water pressure controlled?:
Laboratory effluent—type of system, including neutraliser:
Position of fume cupboards and details of exhaust systems:
Air-conditioning—type of system and distribution; number of air changes; recirculation:
Does overall arrangement provide satisfactory inter-relationship between laboratories and library, stores and workshops?:
How have they solved the problem of storing laboratory chemicals, equipment, test specimens, research data and correspondence?:
Hazardous chemicals—maximum quantity permitted in one laboratory, and where kept. Where are secondary and bulk supplies stored?:
Flammable solvents—maximum quantity permitted in one laboratory, and where kept. Where are secondary and bulk supplies stored?:
What special efforts have been made to simplify housekeeping and maintenance?:
Any other item of particular interest:

SELECTION OF SITE

The success of a laboratory project depends to a very large extent on a wise selection of site. Gone are the days when a site can be chosen because it 'looks good' and the size is 'about right'. Site selection calls for a great deal of consideration, not only of the relatively straightforward technical aspects, but of intangibles such as the amount of space to be provided for expansion, and whether the location will attract scientific and supporting staff. One must acknowledge the remarkable foresight shown in the choice of some sites, or in others perhaps a measure of good fortune! As a general rule, however, the selection of a site does not receive the attention it warrants.

Research laboratories are built at high cost and few government or private organisations can afford a rebuilding cycle of much less than fifty years; this period, it so happens, can be regarded as the useful life of even a well-designed laboratory. In any case, by this time the site is either filled to capacity or many of the buildings are so scattered that efficient operation is impossible.

The project officer should consult the senior scientist concerned, investigate suitable sites and submit a detailed report and recommendations to the management. This procedure will increase the chances of a wise selection and reduce the risk of repeating mistakes made in the selection of other sites. When, as is usually the case, several sites are being considered, the comparison can be simplified by the use of a points score system.

Remember that there is no initial economy so deceptive as low-cost acreage which is too steep, requires special foundations or is remote from services. Whilst it might seem obvious, it must be emphasised that a technical assessment of the proposed site is an essential factor in establishing its overall suitability. Points to be considered will vary, but they should include the following:

Slope—a gently sloping site is preferable; it simplifies drainage and makes for more interesting architectural composition and attractive landscaping. Nowadays earth moving is one of the less expensive building costs; therefore land having a slope which hitherto might have been considered excessive can be converted into a suitable site.

Soil type—test drillings should be made to determine soil types and conditions which might affect the design of foundations. If piles have to be used, then the cost should be known. The presence of rock will mean costly excavating if basements or special buildings are required. *Availability of services*—is an adequate supply of water and electricity available? What is the variation in pressure and voltage? What is the record of service interruptions? What is the position regarding stormwater drainage, sewerage and roads?

In addition to these aspects, information regarding the direction and velocity of the wind, amount of cloud, temperature, humidity and rainfall should be sought from the Bureau of Meteorology; these can have an important bearing on the master planning of laboratory sites. Enquiries should also be made regarding the availability of public transport, proximity to the university and technical schools, and the risk of air pollution, noise and vibration. It is wise to check with the local town-planning authorities to make sure that they are not toying with the idea of building a six-lane highway across or adjacent to the site!

What area is required? There are a number of organisations in the United States of America with quite large laboratory sites; among these are Bell Telephone at Holmdel, NJ, with 460 acres, the National Bureau of Standards at Gaithersburg, Md. (*see* p. 129), with 550 acres and the Bethlehem Steel Corporation at Bethlehem, Pa. (*see* p. 126), with 1070 acres. Whilst it is neither necessary nor practicable for all laboratory sites to be measured in terms of hundreds of acres, nevertheless it is an unfortunate fact that too many sites are far too small. So often it seems that management—because of lack of money, lack of foresight or maybe just lack of thought—will not purchase even a few acres in excess of immediate requirements.

The area is sometimes considered in relation to a predetermined maximum number of people to be accommodated on the site; for example, can a site population of about 1000 be regarded as ideal for overall efficiency, site communication, car parking, etc.? It is not easy to establish such a figure. The staff on the National Bureau of Standards' site already numbers nearly 3000 and, with the considerable provision for new buildings, this number will be greatly increased. Bell Telephone considered their site at Murray Hill, NJ to be at a maximum desirable capacity when the staff numbers reached 4600 (car parking 3000), although there still was vacant land on their 200-acre site.

What are the likely building requirements? Experience indicates that it is almost impossible to make an accurate assessment of needs five years hence, let alone ten or twenty-five years ahead. No matter how much 'built-in' expansion is provided in buildings being planned—for example, under-capacity and unallotted rooms—within five years of occupation this usually proves to be insufficient. Even when what appears to be ample provision is made for extensions to buildings, and space is allowed for new buildings, this very soon becomes inadequate. It might be difficult to imagine requirements doubling in the foreseeable future, whereas they might more than treble. Provision for expansion cannot, therefore, be made on any rational basis; the only way is to cater for the likely and then make a generous additional allowance for the unforeseen.

What will happen if the site is too small? In an attempt to avoid continued development of sites which in the very near future will become too small, existing sites should be reviewed at, say, ten-year intervals. If it appears likely that the site will reach maximum capacity within the foreseeable future, then money spent on the erection of more buildings on this site will be largely wasted; it is much more sensible to face up to the situation and obtain a larger site. Sometimes it is possible to evade this problem because rapid growth has provided an opportunity to subdivide the field of research and thus establish a second unit on another site.

If the laboratories are to be transferred to a new site, financial considerations sometimes dictate that this be done in several stages. This, of course, means temporary separation of staff and activities, and some scientists would throw up their hands in horror at the idea. Whilst not advocating it as a desirable state of affairs, this disadvantage is usually not as serious as it at first appears, and certainly any inconvenience involved is outweighed by the advantage of having embarked on a long-range site development plan.

BUILDING BRIEF

In the design of a complex building there are many hundreds of items to be considered; it is not surprising that some are either overlooked or misinterpreted. The building brief reduces this possibility and serves the very useful purpose of forcing the client not only to get his thoughts in order, but also to record them. To the designing architect and engineer, it is an essential document.

The brief must be accurate, clear and complete. Preparation of detailed information at the commencement of the project will achieve much more satisfactory results than snap decisions made during discussions—sometimes when plans and specifications are nearing completion. Some of the items given in the brief amount to instructions, others a preference or suggestion, whilst still others are included just to make sure that they are at least considered. Precise information regarding such matters as laboratory layouts and numbers of service outlets cannot be given until the scheme is more fully developed.

Primarily, then, the brief is intended as a basis for discussion; it is certainly not meant that a preference expressed by the client should be accepted by the designing architect or engineer if a better suggestion can be offered; on the other hand, no item should be altered or disregarded without the approval of the client. Because of the variations from job to job, it is not possible to establish a standard format for the building brief, but the following is a guide to the type of information which should be included:

Extent of work
Scope of contract
Funds available
Reference drawings, including development of site and orientation of buildings
Structural

- Preferred type of construction
- Column centres and module
- Floor-to-ceiling height
- Floor loading for laboratories, offices and areas of special loading—for example, libraries and stores. Future partitions will be specified as lightweight and only at modular centres. Also, the possible future use of mobile compacted shelving in certain locations should be indicated
- Future extensions
- Details and locations of mechanical equipment, tanks, etc.
- Special requirements
    - Columns not to project into laboratories along external walls or into corridors

In laboratories, holes in floor to have metal sleeves cast in slab and projecting 1 in above finished floor
Isolated slabs
Elevator
Preferred type, size and minimum width of door opening

Architectural

Windows
Preferred type and total width of glass per module
Sill height in laboratories and in offices
Sun protection—extent and type
Partitions
Preferred type
If glazed or double-glazed, give details
Doors
Preferred width and finish
If glass panels, give details
Door furniture
Type and height
Finishes to floors, walls and ceilings
Generally to be . . .
Special requirements—for example, entrance foyer, executive suite, typists, library, conference, canteen, radioisotope laboratories, stores
Laboratory fittings
Drawings showing the positions of benches, fume cupboards, etc., and all service outlets
Details of materials, construction, and finish
Conference room
Capacity
Dais, lectern, switches for lights, microphone
Writing and notice boards, projection screen
Hat and coat pegs
Storage of chairs
Canteen
Capacity
Type of food service
Storage of food, tables and chairs
Notice board
Toilets
Number of males and females per floor
Showers
At least one male and one female per building
Cleaners' closets on each floor
Miscellaneous
Insulation
Position of downpipes

- Flagpole
- Name of building, type of sign and location
- Entrance foyer—directory and notice board, display cabinets, mural

Telephones
- Type of equipment and capacity, provision for expansion
- Position of cable ducts

Roads, stormwater drainage, and car-parking—including provision for extension

Landscaping and fencing

Mechanical

Air-conditioning
- Extent and type of installation
- Temperature, relative humidity and acceptable variations
- Noise level to be a reasonable minimum
- Recirculation from offices and exhaust from laboratories
- Heat load per module
  - Number of persons
  - Equipment
  - Additional capacity in ducts, equipment and switchboard for possible increases in heat loads
- Special requirements
  - System to operate . . . hours per week, excepting hot, cold and constant-temperature rooms which will operate continuously, and certain laboratories which will operate outside normal working hours
  - Relative humidity not to exceed . . . % in certain laboratories—for example, infra-red spectrophotometry not to exceed 45%
  - Independent air exhausts from certain laboratories—for example, chromatography and tissue-culture laboratories
  - Performance of air filters in certain laboratories
  - Air movement at bench level not to exceed . . . ft/min in certain laboratories
  - Position of equipment—for example, fan coil units in relation to laboratory benches, and library bookshelves
- Position of air inlet in relation to fume cupboard exhausts and direction of prevailing wind
- Hot, cold and constant-temperature rooms
  - Preferred type of construction and insulation
  - Temperature and relative humidity, with maximum variation at specified position, and method of testing
  - Heat load—number of persons and hours occupied, equipment, average door entry
  - Fresh air requirements, if any

Ventilation
- Fume cupboards
  - Ducts—position and materials
  - Discharge to be vertical—individual or grouped, height of outlet
  - Fans—position and materials, details of motors

- Special requirements
  - Exhaust of fumes from special equipment, toxic chemical stores and ovens
- Laboratory services
  - Taps—design, type and finish
  - Isolating valves
  - Cold water
    - Softening, filtering, storage and pressure control; give details if required
    - Number of outlets and estimated use per day
    - Special requirements—for example, thermostatic mixing valve and . . . type filter pad in photography
  - Hot water—number of outlets and estimated use per day
  - Chilled water, Burner gas, Compressed air, Vacuum, Special gases, Distilled water, Demineralised water } Number of outlets, extent of reticulation, suggested type, pressure and capacity of plant
  - Laboratory waste
    - Pipe and fittings—preferred materials
    - Neutraliser—position and type
    - If radioactive waste, give positions (to be connected directly to interceptor trap)
- Fire protection (automatic)
  - Suggested type and extent of installation
  - Special requirements—for example, for overnight laboratories and flammable materials stores

Electrical

- Supply
  - Underground or overhead
- Substation
  - Suggested position and capacity
- Additional capacity for expansion in substation, main and sub-boards
- Lighting
  - Generally to be fluorescent, to give . . . lumens per square foot
  - Switches
    - Grouping and position
    - Check provision made for night cleaning and late departure from building
  - Special requirements—for example, photography
  - Security and car-park lighting
- Power
  - In laboratories—type of reticulation, number of outlets (with pilot lights) and mounting height
  - Special requirements—for example 15 A three-phase and, in overnight laboratories, flameproof outlets

PREPARATION OF DRAWINGS

It is preferable for the project officer to make the drawings after discussion with the scientists concerned, rather than having to base them on a number of unrelated sketches prepared by individual scientists. Whilst each scientist knows his own requirements, it is the project officer who knows the scheme as a whole; he can, if necessary, suggest improved alternative positions for fume cupboards, sinks and service outlets which will conform to the general pattern of service reticulation and ducts.

One quite effective method of collating information is for the scientists to prepare a series of cards, each one representing a room and listing its function, size, special fittings and required proximity to other laboratories and shared facilities. These cards can be laid out and rearranged until a satisfactory layout has been developed.

Drawings showing the layouts of laboratories, details of benches and fume cupboards, and the position of service outlets should include elevations of each wall; it is important that *all* service outlets on any particular bench be shown on the one sheet, otherwise mistakes usually occur—for example, a power outlet behind a water tap. Unless the architect is a specialist in laboratory design, it is better that these drawings be done by the client's Building Section.

SITE SUPERVISION

Site supervision of the architectural, mechanical and electrical work is normally dealt with by the supervisors responsible to the designing architect and designing engineer. The project officer can assist in many ways by attending site meetings which should be held regularly from start to finish of construction, and by inspecting the work frequently during progress and pointing out items which require attention; he should also be readily available to answer queries on the many day-to-day problems which arise.

The mechanical and electrical services are the life-blood of any laboratory. They must function with an absolute minimum of interruption otherwise experiments in progress can be ruined, causing loss of much time and money. Supervision of their installation requires meticulous attention—the sort of attention one would expect from the person who will be responsible for their maintenance after completion of the building. The ideal arrangement is for the client to appoint a maintenance engineer as soon as a new job commences. This person can assist the project officer during construction of the building and, depending on the size of the job, he can be entirely responsible for the supervision of the services or he can work with the consulting or supervising engineer. If it is not practicable to appoint a maintenance engineer, then at least someone from the central maintenance shop should make frequent inspections and be present when:

(a) the pipe systems are being pressure-tested, and the mechanical equipment and automatic controls are being checked and adjusted;
(b) the air-conditioning system is being balanced;
(c) the fume cupboards are being tested; and
(d) when temperature and humidity readings are being taken in the constant-temperature and cold rooms—to check that, at the specified conditions and height, they are within the allowable tolerances.

Maintenance can be made much easier if proper attention is also given to seemingly small items—for example, isolating valves should be positioned so that it is possible to repair or, if necessary, remove them. The removal of a nut in an awkward position is difficult in any case, but it becomes almost impossible if the bolt is too long and it is rusted. Most specifications require that bolts do not project more than two threads beyond the nut and that anti-seize compound be used.

In order to avoid trouble with future buildings, maintenance and road making, it is essential that the precise location of underground service mains be known. It is most important that there is also an accurate record of the positions of service pipes and cables, isolating valves and switches around the site and in the buildings, and access panels to adjusting dampers in ducts, etc. This is another job best done by the maintenance engineer who can obtain the information methodically during progress of the job, and record it clearly on a plan with dimensions and explanatory notes. To the uninitiated, this might seem an obvious requirement; in fact, it is rarely done, and the lack of this precise information causes unnecessary trouble later.

CONTINUITY OF DESIGN AND CONSTRUCTION

A survey of research laboratories built in the last decade indicates a wide variation in design. Room sizes differ greatly, so do the floor-to-ceiling heights and the widths and extents of corridors; even the design of laboratory fittings varies. In some laboratories flexibility has been carried to an extreme, and in others it has not been considered at all. It follows that there is a wide variation in costs.

It is a mistake to assume that above-average cost produces a more efficient laboratory and greater satisfaction to the scientist. Few, if any, laboratories would stand a detailed cost/value scrutiny without revealing both mistakes and extravagances. Some architects have a cost-conscious approach to design, but there are others who give scant attention to this aspect; they are more concerned with aesthetics than the mundane matter of money.

The rapid development in building materials and techniques makes it virtually impossible for architects to keep up to date with all costs. Quantity surveyors and some building contractors have this information, but unfortunately they are rarely invited to critically examine designs and detailed drawings from the value-for-money point of view. In the best interests of the client there should be cost/value consultations between architect and quantity surveyor or building contractor right from the early stages of design.

A few large organisations have realised the advantage of recording details regarding the design and construction of their laboratories. One notable example is Johnson & Johnson, New Brunswick, NJ, USA, who made a survey of this subject; the 13-man study group was led by their Director of Construction, Mr C. D. Huber. The introduction to their comprehensive and most valuable report states:

> 'Readers of this report must clearly understand its objectives and purposes in order that it not be misinterpreted as a firm recommendation for the exact methods to be used. The watchword of this report is 'ECONOMY'—not what is superlative or especially desirable—not what is merely cheapest, but what is most economical in relation to the purpose intended.
>
> 'As everyone concerned with construction and design knows, unnecessary costs have been

incurred in past buildings because of convention, habit, decisions by uninformed or unqualified personnel, or because of emotional architectural or engineering decisions. The record of such costs has been impartially examined, and recommendations made to avoid them in the future.

'Some of the recommendations in this report may seem drastic, unconventional, or even petty in their scope. It must be remembered, however, that each and every building material and method, system and piece of equipment is a potential source of savings—varying from structure to structure in its scope. The addition of numerous small savings can often result in an appreciable total, and the smallest source of savings must not be overlooked.'

This is not the place to discuss the merits and demerits of inviting public tenders, but an organisation is unlikely to get better laboratories if they have a different contractor for each building. The chances of engaging the same contractor are improved if selected tenders are invited from a group of builders chosen for their high standard of workmanship, organisation and reliability, and for their financial stability. The same principle applies to the supply and installation of laboratory fittings; experience has proved that, to get the best job, this work should be done by firms who specialise in their construction. Provision should therefore be made in the contract for this work to be handled by a nominated subcontractor, or for the fittings to be supplied by the client.

In large organisations most laboratories are built on a 'one off' basis. There are often unnecessary variations in the client's requirements for each building, and a different architect is engaged; to complete the picture, there is a different contractor and a different clerk of works. This procedure is wasteful and does little to prevent mistakes being made or, even worse, the same mistakes being repeated. The aim should be to determine those features of laboratory design which have proved satisfactory and insist on their adoption. It is equally important to decide on unsatisfactory items and to make sure that they are not repeated. If change is really necessary, then there is a much better chance of improvement if an existing detail is re-examined, rather than starting afresh. The best way to combat extravagances and mistakes in laboratory design and construction is to establish an efficient team consisting of project officer, architect, engineer, building contractor and clerk of works. Once this group is established, it is a tremendous advantage if they can work together on future projects. One of the obvious benefits of having a Building Section is that it provides continuity of laboratory design information, and continuing liaison with the designing architect and engineer.

Chapter 2

Laboratory Planning

Laboratory planning is generally regarded as one of the most difficult assignments with which an architect can be confronted. It involves the development of a layout to meet an exacting set of conditions, and the integration of complicated engineering services.

It is essential that the module and layout of the individual laboratories be considered in detail before even preliminary sketch plans are prepared. This can best be done in the following sequence.

MODULE

A module of 10 ft is recommended; this is the distance from centre to centre of two peninsular benches, and it is based on a bench width of 5 ft with a space of 5 ft between. In a one-module laboratory it is the distance between the centre of one partition and the centre of the next; it is based on a wall thickness of 4 in, a bench 2 ft 3 in wide on one side and a table 2 ft 6 in wide on the other—to give a space between of 4 ft 11 in. Generally an entirely satisfactory and clean-cut layout can be planned with the 10 ft module, but if it is necessary to have greater flexibility (*i.e.* rooms 15 and 25 ft wide), then a module of 5 ft must be used. Of course, the module is dependent on the width of the benches and the space between them. The most convenient metric equivalent is a three-metre module.

Width of bench

In chemistry laboratories, the generally accepted width of benches fitted with reagent shelves is 2 ft 6 in for wall benches and 5 ft for peninsular benches. In physics laboratories, widths of 3 ft and 6 ft are sometimes preferred, with a wide shelf for electronic equipment. In some laboratories, a bench width of 2 ft or 2 ft 3 in is adequate. Where solid timber tops are used, the consideration of width in relation to cost is relatively unimportant, but where sheets of some material are being used, the width should be considered in relation to sheet size so that waste is reduced to a minimum.

Space between benches

As building costs rise, it is to be expected that the distance between benches will receive closer scrutiny. Some research laboratory planners maintain that the increasing use of mobile equipment justifies the adoption of a 6 ft space. If it is adopted, then in a building 200 ft long it means the loss of one two-module laboratory; conversely, a decrease from 5 ft to 4 ft 6 in means a gain of one one-module laboratory.

The distance should be determined by considerations of convenience and safety, *i.e.* one person should be able to pass another (working at the bench) comfortably and without risk of collision if the latter should step back unexpectedly. Experience has shown that 4 ft 6 in to 5 ft is ideal; 4 ft is cramped. It must be admitted that there are laboratories in which one man works between benches separated by as little as 3 ft 3 in, but such a small space should certainly not be thought of when planning a new laboratory.

In student and routine laboratories where there is less bench space per person and often two people will be working back to back immediately opposite each other, the space between the benches should be greater than 5 ft so that there is room for others to walk down the centre.

LAYOUT OF LABORATORY

Having established the module, it is now necessary to settle the size and position of laboratory offices, the depth of laboratories and the position of service laboratories, fume cupboards and service ducts. All of these are vitally important in themselves, and of course they actually determine the type of layout which is to be adopted. Let us consider each of these items.

Laboratory offices

There are many scientists still alive today who have worked in laboratories where offices were not provided; the lucky ones had tables in the laboratory and the others just shifted some equipment off the bench to make space for report writing. For a number of years now, it has been standard practice to provide every scientist with an office; it is quite usual to provide individual offices for senior technical officers also, whilst laboratory assistants are expected to share offices or have writing spaces provided for them in the laboratories.

The best location for laboratory offices is always a controversial subject. Are they to be within the laboratory, adjoining the laboratory, on the opposite side of the corridor or grouped in a separate part of the building? Is it essential for all offices to be on an external wall?

Some senior scientists consider an 8 ft by 6 ft office within the laboratory entirely satisfactory. These people spend most of their time actually working in the laboratory and the closeness outweighs the advantages of greater privacy and silence in a larger office across the corridor. In any case, for report writing it is much more satisfactory to use a carrell in the library. The internal office shown in Fig. 1 has a 6 ft by 2 ft 6 in table with bookshelves above and a filing cabinet beneath. This layout has the advantage that the full length of the building is available for laboratories and, with an off-centre corridor, the service laboratories can be conveniently located along the opposite side.

Offices which adjoin laboratories also have the advantage of closeness and they can be larger than the internal office—one dimension is fixed by the module of 10 ft—but they do have the disadvantage that they use the more expensive serviced area. The alternative is to provide offices along the unserviced area on the opposite side of the corridor, but many scientists consider this separation from the laboratory undesirable, and the further the offices are from the laboratories, the more serious this becomes. In the case of offices grouped on another floor, the scientist may even think twice before making the effort to get to his laboratory.

Some scientists consider 10 ft by 10 ft an absolute minimum for an office, and others argue strongly for 10 ft by 12 ft, or even 10 ft by 14 ft. Certainly, when the offices are along one side of a corridor, a depth of 14 ft makes it possible to get a more satisfactory layout for stairs, toilets, etc.

For large projects, it is necessary to consider laboratories on both sides of the corridor; in this case, offices must be either in (or adjoining) laboratories or grouped in a separate wing of the building. For still larger schemes, the double-width layout provides the best solution.

Details of the various positions of offices are shown in Figs. 2, 3, 4 and 5.

Depth of laboratories

Over the last 40 years, the depth of laboratories has increased from about 16 ft to 24 or 25 ft, with some going to 27 and even 30 ft. This has resulted in a better utilisation of space and, as the span is within economic limits, the additional area is obtained at a lower cost per square foot. For the standard type of peninsular bench layout, a clear depth of 24 ft is recommended.

Service laboratories

These laboratories are either planned to be integral with the laboratory and laboratory office unit or they are provided on the opposite side of the corridor; again, the various positions are shown in Figs. 2, 3, 4 and 5. Much of the equipment housed in these rooms is expensive and therefore it must be shared; it follows that this equipment must be located so that it is convenient to the maximum number of staff.

Fume cupboards

The risk of accident is greater in a fume cupboard than elsewhere in the laboratory; so, for reasons of safety, they should not be located where they will block an exit. Of course, if there is an alternative exit from the laboratory, this difficulty does not arise.

Fume cupboards require an exhaust duct with a diameter of from 8 to 12 in. Preferably the duct should connect from the top centre of the cupboard and rise vertically to discharge the fumes above the roof. This does not present a problem in a single-storey building—except, perhaps, when the architect insists on some symmetry in the positions of the outlets on the roof. However, in a three-storey building, the position of the fume cupboards and the space required for exhaust ducts become more involved; if, in the preliminary planning stage, time is spent working out these details, it will obviate later troubles such as horizontal ducts which are too long or riser shafts which are too small.

The installation can be simplified by having the laboratories requiring the most fume cupboards on the top floor; quite often the entire ground floor space can be allotted for rooms and laboratories without any fume cupboards.

Service ducts

The mechanical services are a major feature of any laboratory and, in order to achieve good design, location and accessibility, they must be given a lot of thought. In some laboratories the installation will involve three or four pipes, and in others there might be six or more.

For benches serviced from the external wall, there should be horizontal and vertical ducts with removable covers. For benches serviced from the corridor wall, it is necessary to have a vertical duct accessible from the corridor. In some laboratories—especially if island benches are being used—the service pipes are reticulated in the space between the floor slab and the removable ceiling. This system does have the disadvantage that it requires many holes through the floor and, in the event of floods, these will cause trouble in the room below; also, repairs and alterations seriously disrupt work in the laboratory and, what is worse, it is somebody else's laboratory! Nevertheless, this system is preferable to the use of ducts in the floor because, even at high cost, it is quite difficult to get a cover which is removable, serviceable, rigid, neat in appearance and perfectly flush.

For large projects where the double-width layout has been adopted, a service corridor is the obvious solution because it provides excellent accessibility to horizontal and vertical pipes and, in addition, space for fume cupboard exhaust ducts and miscellaneous laboratory equipment such as pumps.

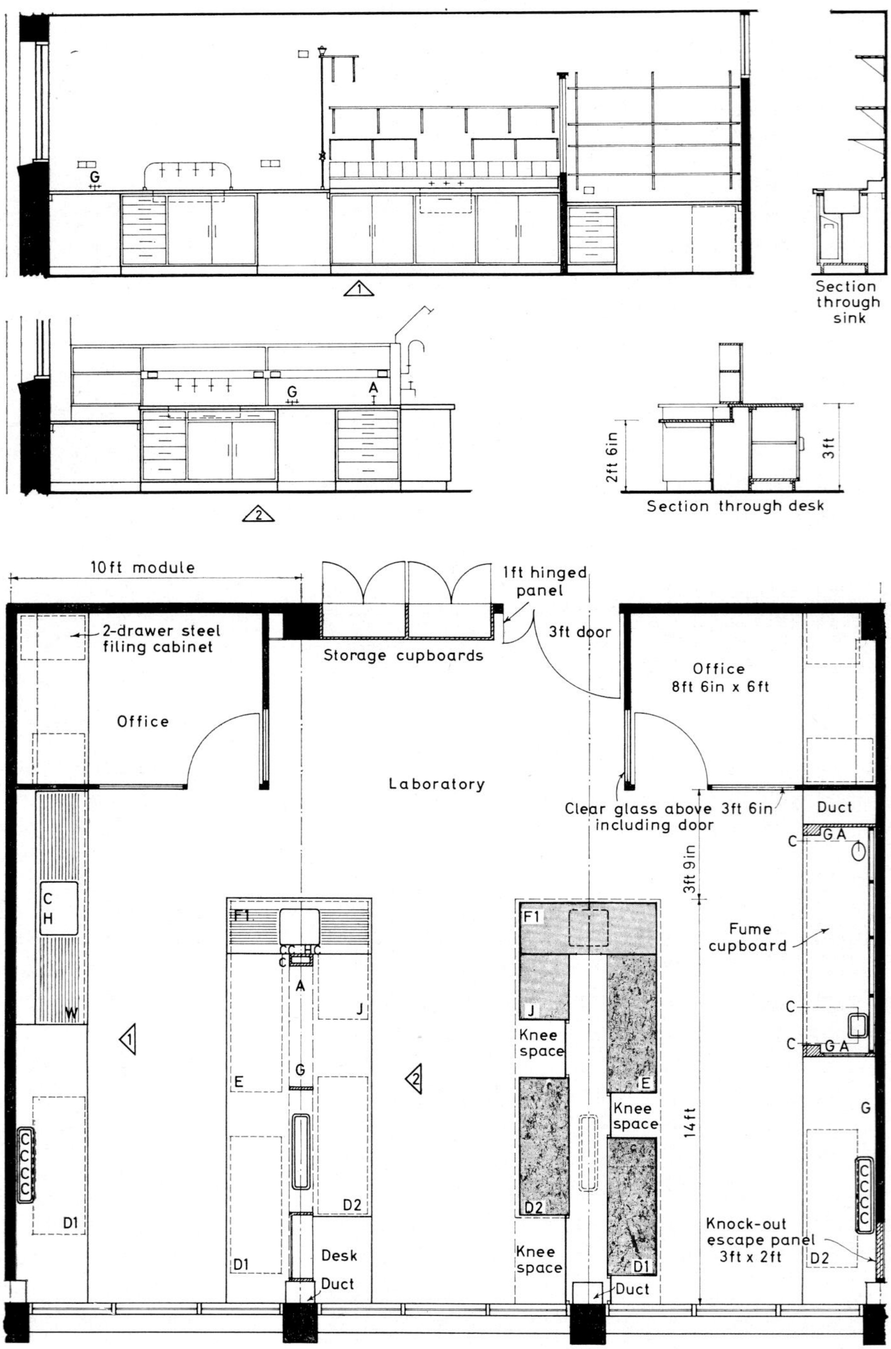

FIGURE 1 *Layout of a three-module 30 ft by 24 ft laboratory.*

Type of bench

There are three types of bench—peninsular, island and wall. As the names imply, the peninsular bench projects from the wall and the island bench is free-standing.

With the greater depth of laboratories, the use of peninsular benches at right angles to the windows has become almost mandatory. They are preferable to island benches because the installation of services is easier and less costly, and there is minimum shadow when they are fitted with reagent shelves. Most laboratory workers will no longer argue that the extra space required to give access to four sides of an island bench is justified.

As a general rule, wall benches under windows should be avoided; facing the sun in front of windows on the east and west elevations makes working conditions quite intolerable. For windows facing north, screening the low-angle sun in the winter is not always satisfactory; even with south-facing windows, glare can be a problem. Wall benches between peninsular benches create inaccessible pockets on either side and, for this reason also, they are not recommended.

Whether it be a one-, two- or three-module laboratory, the combination of peninsular and wall bench at right angles to the external wall produces the simplest layout. The one-module laboratory provides the most wall space per unit area; the three-module laboratory has the widest application because in many cases it accommodates the optimum number of staff to share equipment and facilities.

Details of a layout which has been used quite extensively are shown in Fig. 1. This layout can be adapted to meet a wide range of conditions—for example, one or both of the offices can be omitted, the number and type of bench units and service outlets can be varied, the reagent shelves can be reduced in length or omitted, or one whole bench can be omitted to leave space for equipment or a rig for setting up apparatus.

Prototype laboratory or bench

For large projects, it is a very good idea to have a prototype laboratory, and for small schemes at least a prototype bench. If these are to achieve their real purpose, they should be complete with services and accurate to the smallest detail. Most scientists can read plans very well; however, there are always some who can't visualise the finished product, and for them, and for the builder and his subcontractors, a prototype is a great help. Invariably, after examination and discussion, some improvements or economies are effected. Also, when a prototype is available for inspection by tenderers, its cost can be offset by more accurate estimating.

Windowless laboratories and offices

Given a choice, most people would prefer to work in a laboratory which has windows; it is very pleasant to be able to look out on a garden or landscape, or even to get a glimpse of the sky. There is a prejudice against working in rooms without windows because it is thought that they create a sensation of being confined. The objection to this feeling of lack of contact with the outside world can be partially overcome if it is possible to 'look out if you want to'—for example, in some double-width laboratories, the door to the internal laboratory is opposite the door of the external office, and both are in line with the window; the doors have clear glass top panels. In one windowless laboratory I have visited, I was interested to see a brightly coloured landscape hanging on the wall of an internal office; in another—a physics laboratory—many of the staff have worked quite happily for years in basement rooms; in yet another which has windowless laboratories and offices, the Director told me that, after 12 years occupation, 'early apprehension that a closed-in feeling due to lack of outside windows would be a problem has not materialised'.

Laboratories without windows are shielded from the sun and external temperature variations, and it is possible to get much more accurate temperature control; another asset is more wall space. My impression is that windowless laboratories (and, to a lesser extent, windowless offices) are likely to be accepted more readily in the future.

Width of corridors

Factors which determine the width of corridors include the amount of traffic, the length of the building and whether the doors open in or out; in overseas laboratories it is usual for doors to open into the corridors. Relevant details regarding the five laboratories described in Chapter 8 are:

	Width of corridors	*Length of building*	*Doors*
Abbott	6 ft	224 ft	Single, 3 ft wide, opening out, serving rooms one side only
Battelle	7 ft 6 in	276 ft	Single, 3 ft 3 in wide, opening out
Bethlehem	7 ft	315 ft	Double, 2 ft 10 in + 1 ft wide, opening out from laboratory wall of service shaft at which point the corridor is 12 ft wide
Hoechst	6 ft 6 in	328 ft	Single, 3 ft 3 in wide, opening out from laboratory wall of service shaft at which point the corridor is 9 ft 9 in wide
National Bureau of Standards	7 ft	385 ft	Double, 3 ft + 1 ft 6 in wide, opening out from laboratory wall of service shaft at which point the corridor is 12 ft wide

There is very little traffic in the corridors of research laboratories and in Australia, where the doors generally open into the laboratories, a width of 5 ft 6 in is adequate; furthermore, the narrower width helps to prevent the motley collection of refrigerators and cupboards which so often are lined up along one or both sides of the corridor. Nevertheless, 5 ft 6 in is an absolute minimum and assumes that there are no projecting columns; if the length of the building exceeds 200 ft, this width should be increased slightly to be visually acceptable.

Adoption of a basic laboratory layout

Every effort should be made to develop a basic layout which is standard throughout the building. This is not easy because on every job there is generally at least one scientist who, without any real justification, insists that his office or bench should be in a different position, and he will advance reasons why his idea of layout is necessary for some particular investigation. If he wins his argument and his laboratory layout is non-standard, it so often happens that the project stops, or he leaves, and it is almost certain that his successor will require a different layout. On the other hand, there are some situations where it really is necessary to meet particular requirements, but these can and should be met by variations within the basic layout.

The establishment of a basic layout requires some firm decisions by the officer in charge, and he must apply these with a certain amount of ruthlessness if this proves necessary.

LAYOUT OF BUILDING

Whilst the planning of each laboratory building has its individual problems, the range of layouts can be narrowed to a few which have been found satisfactory. Depending on the size of the project, the type of work and the space available, any one of the following can be recommended.

Off-centre corridor

This layout has wide application for relatively small schemes—for instance, from a single-storey building 100 ft long to several two- or three-storey buildings of about 200 ft in length. It has the great advantage that all the laboratories can have a south-facing aspect, and the two room depths provide flexibility in planning. Four variations of this layout are shown in Fig. 2.

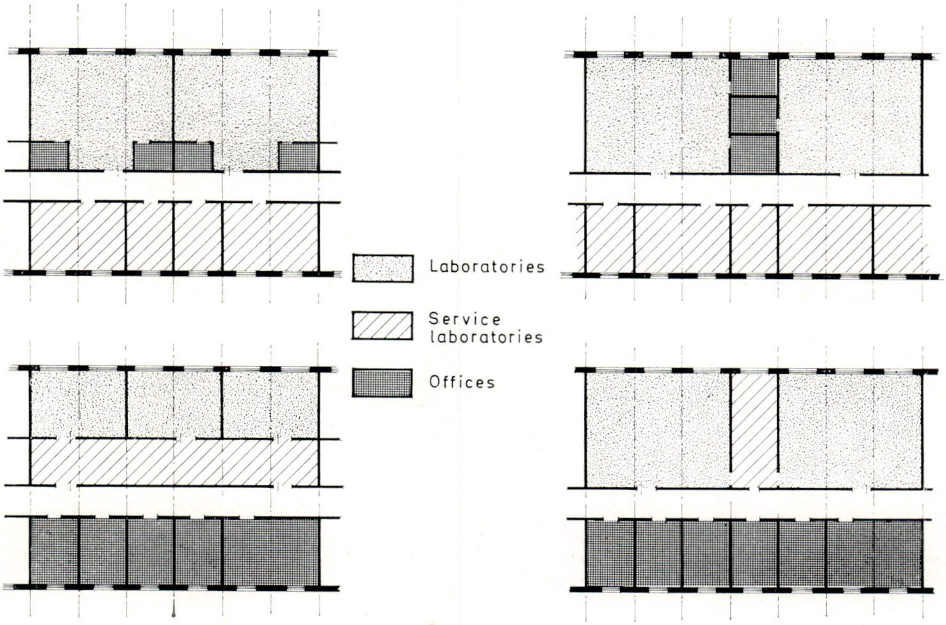

FIGURE 2 *Off-centre corridor layouts, showing the relative positions of laboratories, service laboratories and offices.*

Central corridor

This layout is more suitable for larger schemes. It has the advantage that the grouping of laboratories is more compact because they are on both sides of the corridor. Also, as the same width corridor is serving a wider building than in the case of the off-centre layout, it provides a greater assignable space. However, it does mean that half the laboratories have a north-facing aspect. Two variations of this layout are shown in Fig. 3.

Double corridor

This layout provides a good interrelationship between laboratory, laboratory office and service laboratory, and it sometimes offers the best solution when the width of the building is fixed within certain limits. It has the advantage that, as the service laboratories are windowless, it is easier to obtain accurate temperature control; in many cases, the absence of natural light is an asset. (*See* Fig. 4.)

Service corridor

The double-width layout shown in Fig. 5 is especially suitable for large schemes. As laboratory services become more complex, and temperature control more critical, it is likely that this type

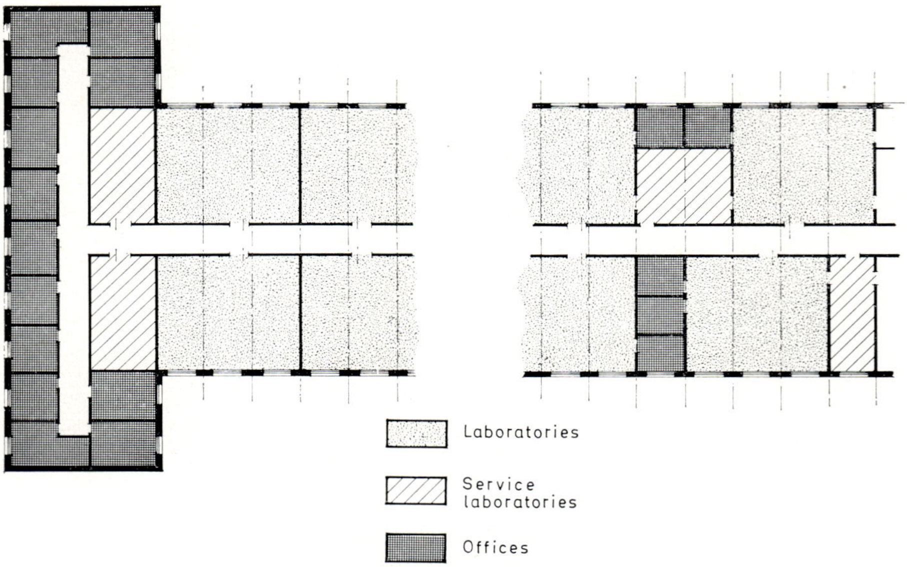

FIGURE 3 *Central corridor layouts, showing the relative positions of laboratories, service laboratories and offices.*

of layout will be more widely accepted. The increased area at one level contributes to more efficient operation because the scientific staff are brought closer together and the sharing of equipment is facilitated.

Assignable area

The gross area is the overall area of the building, the assignable area is the actual area of usable space, and the difference is the combined area of entrance halls, corridors, stairs, toilets, ducts and wall thicknesses. The 'use factor' is the ratio of assignable area to gross area, and it ranges from approximately 50 to 70%.

The best utilisation of space is obtained by having one corridor serving rooms on both sides. For example, in the simplest type of three-storey building with minimum entrance hall, a 5 ft 6 in corridor with 24 ft deep laboratories along one side and 14 ft deep service laboratories along the other:

Gross area	201 × 47		9447 ft^2
Assignable area			
Laboratories	196 × 24	4704	6762 ft^2
Service laboratories (allowing five modules for stairs, toilets, elevators and ducts)	147 × 14	2058	

$$\text{Use factor} = \frac{6762}{9447} = 71\%$$

Obviously, a corridor serving rooms on one side only, or two corridors serving three rooms, decreases the ratio of assignable to gross area and therefore increases the cost.

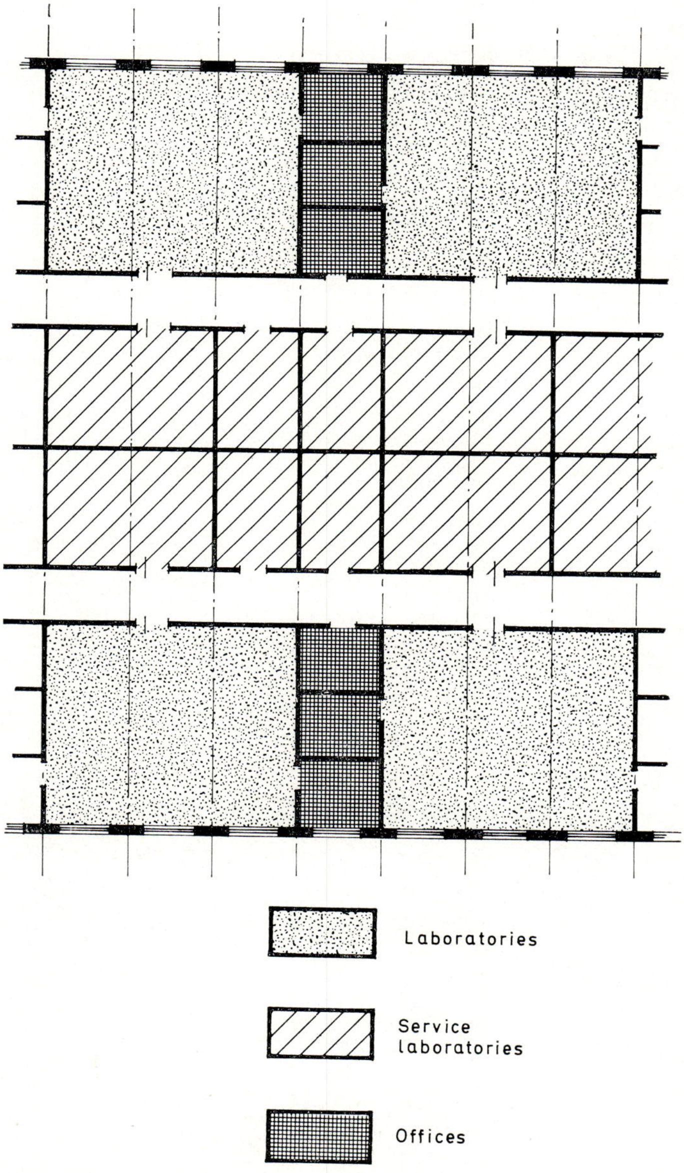

FIGURE 4 *Double corridor layout, showing relative positions of laboratories, service laboratories and offices.*

Floor space per person

The space required by scientists varies greatly. Most require a laboratory, a laboratory office and access to several service laboratories, but quite a number need additional facilities such as glass-houses, animal pens or large areas for pilot-plant investigations or the preparation and storage of many hundreds of specimens. Then again, some scientists use equipment which is small and

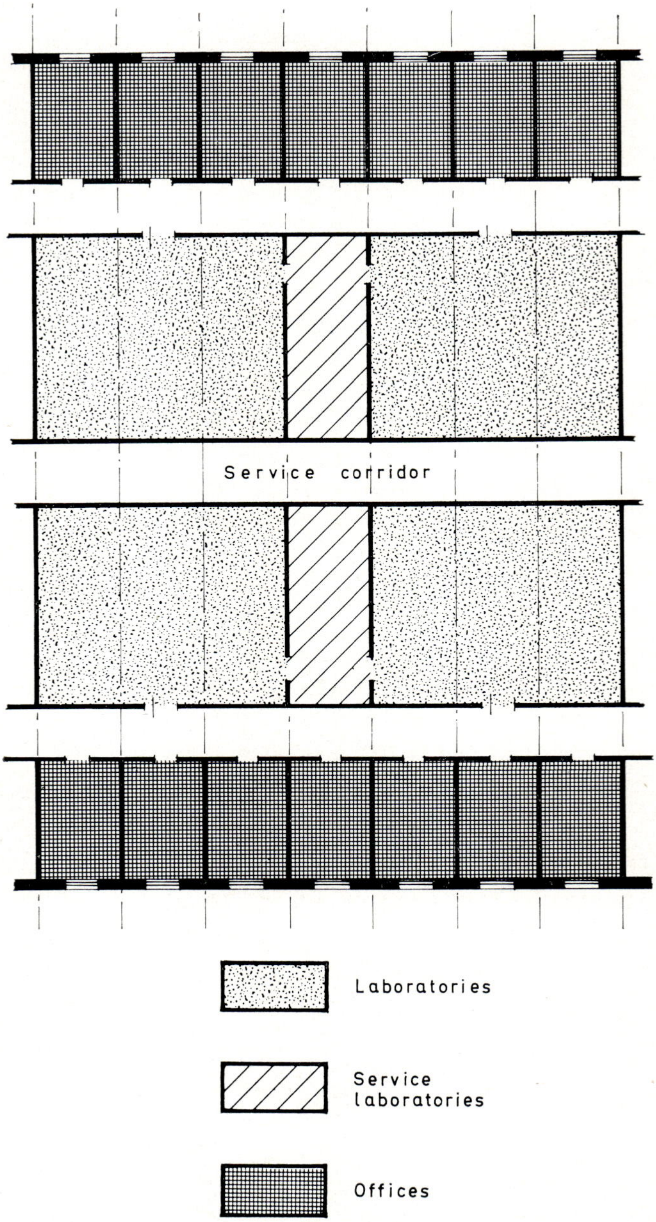

FIGURE 5 *Double-width layout with service corridor, showing relative positions of laboratories, service laboratories and offices.*

commercially available, whilst others must have large equipment which often has to be specially designed and fabricated in workshops on the site.

In his report 'Space Requirements for Scientific Research Laboratories', J. Lorne Gray states:

> 'The results of a detailed study of space occupancy in the National Research Laboratories, Canada, are offered as a help to those who are designing new laboratories or appraising the degree of utilisation of existing laboratories. . . . Only those laboratories are reported which are occupied to maximum capacity or near-maximum capacity, and as there are many laboratories of different sizes and character reported in the study, and as the total number of employees, 1900, is sizeable, the overall figures found can be considered representative for fully occupied laboratory buildings of the general type under consideration. . . . The information obtained from this study suggests that in planning ordinary biological, chemical, physical, and engineering research laboratories, 350 square feet gross area per person should be considered a minimum. In order to allow for a reasonable increase of staff and flexibility in operation, the National Research Council has adopted a figure of 400 square feet of gross space per person as a design specification for new laboratories of this type, with an estimated use factor of 70 per cent. These figures are not applicable to one or more detached laboratory rooms but only to a building or group of buildings in which are housed all the administrative, office, shop, and store space needed for the operation of the laboratories.'

This report was published in 1949 and, in order to compare these findings with accepted CSIRO practice more than 20 years later, an analysis of 12 laboratories was made; with one exception, they were all occupied at near capacity. In eight of the laboratories, the gross area per person (averaged over the total staff) ranged evenly between 370 and 440 ft^2. In the other four laboratories, the range was 510 to 550 ft^2; of these, one had considerable unallotted space, one was a physics laboratory and the other two had large pilot-plant areas. This analysis indicates that, 'in order to allow for a reasonable increase of staff and flexibility in operation', a figure of 450 ft^2 should be used. However, it must be kept in mind that there is equal justification for some laboratories requiring 550 or even 600 ft^2 per person.

It will be seen from the above that it is difficult to apply a predetermined 'space per person' figure and, in fact, it can be misleading. When planning laboratories and ancillary buildings, I have found it much more satisfactory to relate the size of individual rooms to the number of occupants—for example, the three-module 30 ft by 24 ft laboratory shown in Fig. 1 provides ample accommodation for two scientists and two assistants; it represents 180 ft^2 per person, including office space, but not service laboratories. An area of 150 ft^2 per person indicates overcrowding.

Bench space per person

One measure of good laboratory accommodation is adequate bench space. The layout shown in Fig. 1 provides 62 lineal feet of bench; this represents 15 lineal feet per person for four (with a maximum of five) persons. A bench length of 12 to 15 ft per person is a generally accepted standard, and an uninterrupted length is preferable to several short lengths. For an excellent analysis of the length of bench required for low, medium and high disciplines, the reader is referred to Chapter II of 'The Design of Research Laboratories' by the Nuffield Foundation.

Chapter 3

Design of Building

There is almost no limit to the number of items which should be considered when designing a laboratory. This chapter deals with some of the more important aspects.

NUMBER OF STOREYS

In the majority of cases a single-storey building is an extravagant use of land, and therefore a two-storey scheme is preferable; this also results in a more compact layout of laboratories and services, and there is more scope for architectural treatment. Three-storey and occasionally four-storey schemes provide satisfactory and economic solutions; but with more than three storeys, the building cost will increase, and therefore this should only be considered when there is no alternative.

FUTURE EXPANSION

Much time and money can be saved if, in the planning and siting of the building, detailed consideration is given to providing for additional accommodation in the future. The master plan for a large building on a large site should also provide for a second, third or perhaps even more buildings. In smaller schemes, provision can be made for the building to be extended at one or preferably both ends; the latter has the advantage that the existing rooms can be rearranged more easily.

It so often happens that the expansion which takes place during the planning and construction period absorbs most, if not all, of the unallotted or under-capacity rooms provided in the preliminary plans. For this reason it is a sound and economic proposition to have an unfinished semi-basement or to raise the building to provide an open ground floor. This is a much better solution than adding a storey because, apart from reducing construction noise and inconvenience to staff, it eliminates the high cost of shifting mechanical equipment in the roof space at a later date. However, if the site offers no alternative, it is better to accept these disadvantages and provide for an additional storey, rather than erect a building which just cannot be enlarged.

It is not easy to foresee future building requirements, but it is important that a serious attempt to do this is made, and that the information is indicated on the master plan to enable roadways, service mains and stormwater drains to be sited so that they do not interfere with future development.

STRUCTURAL

It is assumed that the distance between column centres is related to the module adopted for the layout of the benches.

Vibration must be kept to a minimum and the simplest way of achieving this is by using heavy construction. Load-bearing brick walls with reinforced concrete floors are the most effective and reinforced concrete frame is preferred to steel. Load-bearing walls are suitable for two- and three-storey construction; they cost appreciably less and are particularly recommended for physics laboratories where the absence of vibration is a major consideration. One suitable method of construction is the reinforced concrete ribbed floor which has the following important advantages:

constant depth gives maximum rigidity and therefore minimum vibration;
the absence of cross beams simplifies pipe runs and ceiling treatment;
holes for pipes can be made beside the columns;
between ribs, the increased height can be just enough for some tall equipment which otherwise could not be accommodated in the laboratory;
there is no structural complication if, at a later date, it is necessary to cut a hole in between the ribs either for tall equipment or a duct;
in countries where steel must be imported and therefore is expensive or in short supply, the depth of the rib can be increased to reduce the amount of steel used.

The rib spacing should be as wide as possible, *i.e.* 2 ft 6 in for a 10 ft module.

The use of exposed aggregate precast concrete offers new possibilities for shapes and finishes of columns, and precast concrete beams can speed construction. However, precast concrete is not recommended for the floor slabs because it is less rigid than a cast *in situ* slab, and it is more difficult to cut holes if this becomes necessary later.

FLOOR LOADING

The Standards Association of Australia Loading Code CA.34 Part 1, 1969 states that laboratory buildings shall be designed for live loads of not less than 60 lb/ft^2.

The live load depends on the usage of the building. Some chemistry laboratories have been designed for 60 lb; others have been designed for 70 lb on the upper floors, with 70 lb for the offices and corridors on the ground floor and 100 lb for the laboratories. Physics laboratories are generally designed for 100 lb and occasionally 150 lb in the laboratories and 70 lb in the offices and corridors. Again, depending on the layout, libraries can be 100 to 150 lb, with stack rooms and rolling stack shelving areas 150 to 200 lb.

In CSIRO we made a restricted survey of loading and the details are given in Appendix 2. This survey involved weighing the bench units in several laboratories—complete with services, chemicals and apparatus. A rolling stack unit in the library store room, filled to capacity, was also weighed.

VIBRATION

With the increasing use of sensitive instruments and the necessity to eliminate or reduce vibration, it is important that consideration be given to the following:

Location of the building on the site to be a maximum distance from possible sources of vibration.

Use of load-bearing or reinforced concrete frame construction.

Compressors, etc. to be in a central plant room in a separate building or, if this is not possible, a maximum distance from sensitive equipment. Pipes should be fitted with flexible connections; these will also reduce the transmission of noise along the pipework.

Vibration is best dealt with at its source by considering the position and type of equipment, increasing the mass of the foundation and isolating it from the main structure, and by using anti-vibration mountings.

A timber floor is not a suitable support for a balance bench; it is too easily vibrated by people walking on it.

Avoid tall slender piers. Wide spacing of piers contributes to the stability of the instrument.

Care must be taken to ensure that there is no physical contact between the foundation or piers and the floor slabs, pipes under the floor and structural footings; they should therefore be separated by a clear space of at least $\frac{1}{2}$ in. Ideally, this space should be left open, but in order to exclude dust it is usual to seal it at the top with a material which is permanently resilient. It is not easy to obtain a material which fulfils this requirement although, in recent years, suitable foam plastic strips have become available—for example, 'Compriband'. The importance of this detail is stressed and it should be closely supervised during construction; if the significance of it is not pointed out to the foreman, he might not adhere strictly to the specification. Many cases could be quoted where the effectiveness of the isolation has been completely destroyed by poor design or the use of a sealing material which has hardened with age.

ROOF SPACE

A roof space is a most desirable feature and it should have access stairs, trusses to give head room, a catwalk and a large manhole with block and tackle above to permit handling of equipment. Ideally, this area (or at least a portion of it) should have a concrete floor slab.

The main items which can be installed in this space are the fume cupboard exhaust fans, air-conditioning equipment and ducts, distilled water units, cold water supply tanks and water treatment plant. Also, as a last resort, some laboratory equipment can be stored in this space.

ELEVATORS

Provision must be made for bringing large equipment into the laboratories. In the past, a cathead and hoist have often been provided for two- and three-storey buildings; at best, this was never really satisfactory. The cost of a goods/passenger elevator is justified—certainly for three-storey and, if possible, two-storey buildings.

The oil–electric hydraulic type of elevator has the advantages of economy in initial cost, less maintenance, smaller plant room (on the ground floor instead of the roof) and less vibration. The passenger traffic in a three-storey laboratory is very light and therefore the slow speed of 80 ft/min is unimportant. The internal dimensions of the car should not be less than 5 ft by 6 ft and a minimum door opening of 4 ft 6 in is recommended.

FLOOR-TO-CEILING HEIGHT

Over a period of years a height of 9 ft has proved to be adequate; for both practical and aesthetic considerations it should be regarded as a minimum acceptable height. In physics laboratories, a height of 10 ft is sometimes necessary, and there is always the possibility that at least one two-storey laboratory will be required in the building for high equipment and distillation columns.

ORIENTATION

In order to obtain the best working conditions and to avoid sun shining on scientific equipment, it is essential that laboratories be correctly orientated with regard to exposure of windows to the sun. This simply means that a building should have north and south facing windows, *i.e.* its long axis should run east and west. Windows should be avoided in the east and west walls or kept to an absolute minimum; in the southern hemisphere north-facing windows (and, to a lesser extent, south-facing) will receive direct sunlight at certain times of the day and year, but this can be reduced quite easily by the use of horizontal and vertical sun breaks. It is not necessary for sun protection to be designed to completely eliminate direct sunlight (at an acute angle) in the early morning and late afternoon.

In laboratory planning, there is no other single decision which is more important than orientation. All laboratories should face south or, where there are laboratories on both sides of the corridor, north and south. In large projects there is no excuse for this not being so; the architectural concept can, and must, accept that laboratory buildings face north and south and, if necessary, less important buildings face east and west. If site restrictions give no alternative to incorrect orientation, then the areas of glass and the design of the sun protection should be very carefully considered.

As orientation is so important, a brief explanation of true north and magnetic north might be helpful: A compass reading gives magnetic north and, to find true north, it is necessary to allow for the magnetic variation—for example, Perth is 4° west, Melbourne is 10° east and Auckland is 17° east; hence, in Melbourne the true north is 10° west of magnetic north. This magnetic variation alters slightly each year; it is interesting to note that when the site for Melbourne was first surveyed in the 1830s, the variation was $7\frac{1}{2}$° east. It must be realised that compass readings can be inaccurate because the compass itself can be in error up to about 3°, the needle can be deviated by local attraction, a magnetic storm in the vicinity can cause an error of as much as 1° and to a lesser extent there are diurnal and yearly variations. This possible inaccuracy is unimportant for most small buildings, but for large buildings or schemes, the position of true north should be determined accurately by making an astronomical observation on the sun.

Details of the principles involved in finding the extent and direction of sunlight and shadows on buildings, together with a series of charts, are given in the Commonwealth Experimental Buildings Station's Technical Study 23—'Sunshine and Shade in Australia' by R. O. Phillips. The extent of shadow at any given time depends on the orientation of the building and the azimuth and altitude of the sun. This can be obtained from solar charts by using a 'shadow-angle' protractor or by using a solarscope—an instrument designed by the CEBS. The solarscope consists of a horizontal platform around which revolves a mirror which reflects the rays from a powerful spotlight on to a model of the building resting on a platform, thus representing the daily movement of the sun.

WINDOWS

The design should not proceed beyond the preliminary stage before there is a clear understanding regarding the size, position and type of window the architect intends using. Too often the elevational treatment is decided without due regard to the effect the windows will have on the working conditions in the laboratory.

Size

Excessive areas of glass can create adverse conditions; the advantages of high intensity natural light and good outlook can be nullified by the discomfort of glare and the increased cost of heating and cooling. An area of 35 ft^2 per 10 ft module is ample. With a floor-to-ceiling height of 9 ft, a sill height of 3 ft 6 in to 3 ft 9 in and the window head 3 to 6 in below the ceiling, the window will be 5 ft high.

Position

To give the best distribution of natural light, the windows should be central in the module; they are then in the ideal position—in line with the space between the benches. You might think this too obvious to mention, but, in fact, I have seen laboratories where the windows were grouped on either side of the columns. Whilst this might make for interesting fenestration, it does not provide acceptable lighting for a laboratory.

Type

In a laboratory which is not air-conditioned, the occupant should be able to operate the opening sash easily. If this is at the bottom of the window, then the incoming air should be deflected upwards; also, if a transome is really necessary, it should be a minimum depth and not at eye level. The size of the opening sash should be sufficient for a person to climb out in an emergency.

In a laboratory which is air-conditioned the requirements are somewhat different. A fixed window costs appreciably less than an opening window; it also has the advantage of less air infiltration and therefore less dust, and the larger panes of glass improve the appearance. However, there should be at least one opening window in each laboratory—for use if an accident occurs or the air-conditioning breaks down.

COLOUR

The appearance of the modern laboratory has been greatly enhanced by the current trend towards lighter and brighter finishes—white ceilings, light pastel-coloured walls and light- to medium-coloured floor coverings. CSIRO timber bench units are of a light ash with a superior quality, high-gloss, clear polyester finish; solid timber bench tops usually have a clear finish. Colour can also be introduced if terrazzo slabs, plastic (melamine) laminate or vinyl-sheeted bench tops are being used. Above-average finishes and light appearance are appreciated by the staff who usually respond by taking a greater interest in the care and general tidiness of the laboratory.

ART IN BUILDINGS

The imaginative application of art can contribute greatly to the attractiveness and general interest of any building, and there is no reason why laboratory buildings should be lacking in this respect.

Probably more than any other, a country which has exploited this idea is Holland where, for many years, contracts for government buildings have provided for up to 1% of the total cost to be spent on some work of art such as a sculpture, a mural or a decorative panel with a quotation. I remember two aphorisms which read, 'Science always was revolutionary and heterodox; it ceases to be so only when asleep' (Sarton) and 'Work, finish, publish' (Faraday).

Happily, this appreciation of aesthetics has become almost universal and today most laboratories —both large and small—enjoy some form of art. Of course, this can be a very expensive business, but not necessarily so. Only recently I read of a building contract from which the planned works of art had to be blue-pencilled; the architect then produced some 'spontaneous and unthought-out designs and sayings' on unfired bricks and the result was some brickwork of much more than usual interest. If the budget won't run to a fountain or a mural, then it is usually possible to borrow some paintings from the local art gallery on a short-term roster system or at least to have some indoor plants.

DESIGN FOR MAINTENANCE

Much useful information can be obtained by critically examining several laboratories after they have been occupied for several years. Here are some of the items it would be as well to check on:

Have the materials used on the exterior mellowed pleasantly or have they become drab and dirty? Were projecting sills and drip moulds used to prevent dirt being washed down the walls? Have the gutters, downpipes, flashings and stormwater drainage been adequate to cope with thunderstorms? Any ugly staining or corrosion?

Have the floor finishes been satisfactory and how has the skirting stood up to the floor-cleaning machines? What is the condition of the stair-treads? It is a good idea to talk with the cleaning contractor about such matters because most times he can offer useful comments —for example, what does he think about the access to exterior and interior windows (particularly windows in the stair well) for cleaning purposes, and is the switching of the lights satisfactory for cleaning at night?

Has it been necessary to repair laboratory waste pipes? Have additional service pipes and equipment been installed? If so, were there any problems which could be overcome in the future?

I would be surprised if any laboratory would get full marks on even the few points I have listed, and there are many more. If this sort of analysis were made conscientiously, the result would be improved details—probably at less cost.

DESIGN FOR THE TROPICS

Traditional features of laboratories in the tropics have been buildings one room wide, verandahs and spacious rooms with high ceilings. Nowadays, with the advent of air-conditioning, there is no need for them to be basically different from those in temperate zones. The principles given in this book therefore apply also to laboratories in tropical regions and, in fact, some aspects become even more important—for example, areas of glass, sun protection, insulation, hoods at entrances and covered ways for protection from heavy rain. An air-conditioned building in the tropics can have a north–south orientation, regardless of the direction of the prevailing wind; previously, the latter was the deciding factor.

CSIRO Tropical Building Research Note No. 10 states:

> 'Despite the usual belief that conditions in the tropics have some extraordinary deteriorating powers, it has been found by studies in various parts of the world that such conditions are only an intensified form of those occurring in the more temperate regions, and that provided adequate attention is paid to precautionary and protective measures, rapid deterioration is avoidable.'

Precautions should be taken against corrosion and attack by termites and a much more realistic attempt should be made to keep dampness out of the building; there should be sub-floor drainage, weathered sills, etc., and plastic film should be used under the concrete slabs and for damp-proof courses and flashings. More attention should be given to details of design and construction so that maintenance is reduced.

In some parts of the tropics the temperature and humidity range is quite small and this simplifies some building problems. In Bangkok the mean monthly temperature varies between 78·1 to 86·2°F and the mean monthly relative humidity varies between 71·8 and 83·2%; thus there is less seasonal movement in timber, there is not the same need for expansion joints in concrete and quite thin coatings of cement render seem to last surprisingly well; air-conditioning is generally designed to operate at 77°F and therefore heat is not necessary, and cooling to this temperature brings the humidity into the comfort range.

GENERAL PLANNING

The determination of space requirements is based on the existing number of staff plus a margin for expansion. In order to do this on a rational basis, it is necessary for the management to tentatively approve the number of additional staff to be appointed during the next, say, five years.

It is sound planning to allow for all rooms to be under-capacity when the building is first occupied. Even so, within a short time this provision is often inadequate and therefore unallotted laboratories should also be included—if possible, one on each floor. To cope with the expansion of administrative offices, buffer rooms should be strategically placed—for example, a seminar room, drawing office or store, all of which are relatively easy to shift when the need arises.

Library

The subject of library planning and equipment is dealt with in a number of excellent publications, one of which is 'Planning Academic and Research Library Buildings' by Keyes D. Metcalf. However, I do wish to mention the following aspects of library design which, unfortunately, rarely receive the attention they deserve.

Expansion

All too often, inadequate provision is made for the expansion of libraries, despite the fact that information regarding likely future requirements is readily available. For example, CSIRO has a central library in East Melbourne and some 65 libraries and book-holding centres at the divisions and field stations throughout Australia. The whole network now has considerably in excess of one million volumes, excluding an enormous number of pamphlets, films, etc. The largest and most important part of the total stock is the collection of nearly 25 000 periodical titles. During 1970 the central library required nearly 600 lineal feet of additional shelf space. In the same period, a small library with a staff of two serving a division with a scientific staff of 50 required approximately 50 lineal feet; this library had 1500 monographs and 300 periodical titles in 1965 and in 1969 it had 2000 monographs and 400 periodical titles; in 1959 the 13 volumes of the periodical entitled Chemical Abstracts issued for the year required 26 in of shelf space and in 1969 the 22 volumes issued for that year required 46 in.

The size of the library unit (including office and workroom) must be determined and expansion beyond this point must be provided for by having a stack room; ideally, this should be directly below the library or adjoining it. The collection of scientific and technical literature is increasing at such a rate that, in the very near future, libraries will be forced to rationalise storage and distribution. Nowadays microphotography can save a lot of space; in a large organisation, it might also be necessary to have a central stack room—remembering that today a book can be delivered to any capital city in Australia within 24 hours.

Appearance

There is no reason why even a small working library should not be made attractive. With little effort and cost, an impression of drabness can be changed to one of brightness and this will have a noticeable effect on the readers and the library staff. The use of colour and also carpet, curtains, one or two paintings and some flowers or indoor plants can improve the appearance tremendously. Shelving is generally steel of standard design, but for all other fittings such as display racks, periodical cabinets, tables and chairs some latitude in design should be permitted—in fact, it should be encouraged.

Absence of noise

Some people claim that they can concentrate even if seated in a busy thoroughfare, but most will admit that they concentrate most effectively in absolute silence and, quite rightly, they expect a library to be quiet.

The noise level in most libraries is too high. Typing is necessary, of course, but the office can be located so that there is the maximum sound isolation and the door should be positioned with this in mind. A brick partition and double-glazing will help, as will felt fitted around the door, acoustic tiles on the ceiling and curtains at the windows. The floor covering should be sound-absorbing; a dramatic reduction in noise and improvement in appearance can be achieved by using carpet; flexible vinyl is an alternative, preferably with an underlay.

Carrells

These individual rooms can be as small as 5 ft by 5 ft and they provide the ideal solution for lengthy reading and report-writing. Again, silence is essential and, in a space as small as this, even noisy air-conditioning can be a distraction.

If partitioned carrells cannot be provided, then at least there should be one or two tables screened from the noise and traffic. In addition, a small area should be provided for casual reading, preferably with comfortable chairs and carpet.

Conference room—canteen

Every research laboratory must have space for staff seminars, lectures and the occasional social function. It is usual for the conference room and canteen to adjoin—quite often with folding doors between so that extra space for either is readily available. For economy, especially in small projects, one room can serve both purposes. Provision should be made for the following:

stackable chairs, with writing-arms, and tables (preferably folding, to make storage less bulky);
a store for tables and some chairs;
sound-absorbing ceiling and blinds or draw-curtains to darken the room;
generous blackboards and bulletin boards; special lighting to illuminate the blackboards, without glare to the speaker or audience;
dais and rostrum; in some instances, the dais should be removable;
air-conditioning designed to operate at either normal or capacity load.

Chemical stores

The store system must operate efficiently. If the scientist has to fill in a form and wait for goods to be delivered, this can be slow and unsatisfactory and he will then start to hoard unnecessary and excessive quantities of chemicals and glassware in his laboratory; from both space and safety angles, this is very bad practice and should be discouraged. If he has free access to the store, much time and patience will be saved; furthermore, he might come across an item particularly suited to his purpose which otherwise he would not know was available. Even in its largest laboratories, CSIRO operates its chemical stores on the 'honour' system and it works very well; there is no serving staff, but the scientists are expected to record a full description of the articles taken—number, size and grade.

The location and size of the store warrant considerable attention. Provision must be made for the delivery of goods and gas cylinders, and there must be space for empty crates and cylinders and convenient access to the elevator.

Metal shelving is preferred to timber because it is demountable and more readily adjusted. The visibility of stock would be much improved if, instead of the usual dark grey colour, the backs and the undersides of the shelves were white, but, to my knowledge, no manufacturer has yet produced these. The finish at present available is not resistant to some acid fumes, so either an improved finish, or timber (with tray under), should be used for shelves where acids are to be stored.

Sectional workshop and glassblowing

A sectional workshop is a worthwhile feature in most laboratories. It need only be small, but should be available to the laboratory staff for the inevitable running repairs and minor alterations which must be made to apparatus at short notice, and it is a convenient place for a glassblowing bench. Of course, a central workshop should also be provided in a separate building.

Shower and first-aid facilities

At least one male and one female shower should be provided in the toilet facilities for the laboratory staff; also a first-aid room, either in the main laboratory or in an adjacent building.

DETAILS AND FINISHES

Prevention of floods

The incidence of floods will vary greatly; a laboratory might function for five years without one flood and then have two in successive weeks. Irrespective of how often they occur, floods always cause considerable inconvenience, irritation and loss of time—not only to the occupants of the laboratory concerned, but also to the people in the room below. Causes, in about equal proportions, are as follows:

increased water pressure during the night;
malfunction of equipment—for example, microswitch, motor or pump;
carelessness—for example, faulty hose and hose connections or a sink outlet which is nearly blocked.

Many quite simple precautions can be taken to avoid—or at least substantially reduce—the risk of flooding and it is a good idea for the scientist to ask himself these questions: Is the tap washer in good order? Is reinforced hosing being used? Are metal clips fitted to connections? Can part of the apparatus stand on a stainless steel sink and drainer? Has the apparatus been carefully checked last thing at night? Has the night-watchman been alerted—either by a pilot light in the corridor near the laboratory door or a sign stating that an overnight experiment is operating and giving the name, address and telephone number of the person to contact in an emergency?

The extent of possible floods can be reduced if all holes through the floor (for example, for service pipes) are surrounded by a metal strip or dwarf wall. A floor waste in every laboratory might seem a good idea; it has been done occasionally, but it is not very practicable because the grate soon clogs with wax and fluff; likewise, a sloping floor presents practical problems. However, it is wise to have one or two large floor wastes in strategic positions on each storey—for example, in toilets, cleaners' closets, dark rooms or stores—to cope with a major flood or the huge quantities of water used to quell an outbreak of fire; better still, have a vacuum pump available on the site.

If it is frequently necessary for experiments to be continued out of normal working hours, it is advisable to have an 'overnight' laboratory on the ground floor. This should have a sloping cement floor with waste, and a 4 in high kerb across the doorway. The inconvenience of operating a communal laboratory and of shifting and setting up apparatus in another room is less serious than the almost complete cessation of work when a flood occurs.

Floor covering

Asbestos vinyl and flexible vinyl tiles are available at about the same cost per square yard. Flexible vinyl is preferred for laboratories because it is more impervious and therefore a little more resistant to chemical attack; it is resistant to acids, but both it and asbestos vinyl are attacked by some solvents. The flexible vinyl wears better and, being more resilient, is a quieter and more comfortable floor to walk on and is less slippery.

Flexible vinyl is also available in 4 ft wide rolls; the thickness generally used for both sheet and tiles is 0·080 in. The sheet is laid with welded joints; it can be coved around a preformed PVC strip at the junction of floor and walls, and carried up the wall, toe-space or duct to form a skirting; it can even be extended to dado height. In chemistry laboratories, this makes alterations a little more difficult, but it does have the very great advantage that, if a flood occurs, the floor is virtually a tank—provided, of course, that there are no gaps in the skirting around the ducts. Flexible sheet vinyl—with flush welded joints and coved at angles—is an ideal material for use in any laboratory where there is a special requirement for cleanliness and absence of crevices—for example, isotope laboratories.

An attractive floor will contribute much to the appearance of the main entrance foyer and therefore it generally receives special treatment—parquetry, cork, terrazzo or vinyl with perhaps a rug. Carpet is recommended for the Director's suite and, if possible, for the library.

The traffic in the corridors of research laboratories is not heavy; nevertheless, it is desirable to reduce the noise level to a minimum by using a resilient floor covering. If flexible vinyl is considered to be too noisy, then a sponge or cork underlay will reduce this, but it must be remembered that there is a greater risk of damage by indentation if a soft underlay is used.

Partitions

In load-bearing construction, one of the corridor walls and some of the cross-walls are structural, but in framed construction, all internal walls are non-load-bearing. In order of importance, the main requirements for partitions are that they be fire-resistant, sound-proof and demountable. Brick and concrete block (plastered or unplastered), and timber frame sheeted both sides with flush-jointed $\frac{1}{2}$ in thick fibrous plaster or $\frac{5}{8}$ in plaster board, all have a 1-hr fire rating and are commonly used. Unfortunately, walls in chemistry laboratories are quite often sheeted with combustible materials; this is poor practice and should be avoided.

Demountable partitions

Generally, the fire-resisting and sound-proofing qualities are poorer than with conventional construction and the initial cost is higher. They have not been used extensively in Australian laboratory buildings. Whilst they can be used to advantage for laboratory offices, alternate walls between instrument laboratories and the subdivision of administrative space, there is no real benefit in using them for laboratories. Altering the size of laboratories doesn't actually happen as often as might be supposed, but if a wall does have to be moved, the cost is small compared with the cost of shifting the service pipes and benches.

Partitions to laboratory offices

These are very much a matter of personal preference. Although most are glazed above 3 ft 6 in, some people are quite happy with a 5 ft screen wall, others like a door-height partition and others again want a full-height partition.

Doors

There are times when a door opening of 3 ft is just not sufficient to admit a new piece of equipment; at present, 4 ft is considered to be the minimum acceptable width. Rather than waste valuable wall space by having double doors which are unnecessarily wide, the use of a 2 ft 10 in door plus a 1 ft 2 in hinged panel is recommended for all instrument rooms and large laboratories. A clear-glass viewing panel should be provided, varying in size from 6 in by 12 in to about 2 ft 3 in square.

Master keys

It is standard practice—and certainly recommended—that the few keys to internal doors, and keys to all the external doors, be part of a master key system for the whole site.

Writing boards

Writing boards are standard equipment in conference and seminar rooms, and in recent years they have also been used in laboratories and laboratory offices. The common 'blackboard' (sometimes green) is still used, but white Perspex is becoming more popular.

Bulletin boards

Again, bulletin boards are standard equipment in conference and seminar rooms, and in some laboratories. These are generally a special $\frac{1}{4}$ in thick cork board, glued to plywood or particle board backing; alternatively, they can be $\frac{1}{2}$ in thick insulating board, either painted or covered with an attractive fabric.

Shelving

Sagging shelving is unsightly. This small but conspicuous defect is seen too often in laboratories, but it can easily be avoided by a little more attention to the thickness and span of the shelf.

Chapter 4

Laboratory Fittings

There are many good reasons why laboratory fittings of standard design should be used. The scientist can indicate his requirements more quickly and the architect has fewer drawings and specifications to prepare; the savings in cost of manufacture and installation are considerable and the units can be re-used if it becomes necessary to alter the layout of the laboratory.

In Australia, all bench units are timber. In the United States of America, where there is a very large market and the labour costs are high, most are steel; this material is more adaptable to mass production and the units are demountable. They are fire-proof and termite-proof and the maximum precautions are taken to ensure that they will resist corrosion and chemical attack. Before assembly, the steel components are cleaned and treated in a phosphate bath; they are then coated with epoxy resin and heat cured. However, there are still some laboratory people in the USA who are prepared to pay a little extra for timber units.

In 1946 CSIRO designed a range of standard laboratory fittings, and many thousands of these have been used by the Organization and by government departments and private industry. One local manufacturer specialises in the production of the bench units and, in my opinion, they are at least equal to any I have seen overseas. Despite the fact that these units are available commercially, some people still prefer their own design. This results in increased cost, especially when more expensive timbers and unnecessary elaboration are specified.

BENCHES

The various types of CSIRO standard bench units are shown in Figs. 6, 7 and 8. The width of the larger units is 4 ft 9 in which was determined by the width at the end of a peninsular bench, *i.e.* 5 ft less $1\frac{1}{2}$ in bench top projection on both sides; incidentally, this size is about the maximum which can be conveniently man-handled on the job. To provide flexibility in planning, it was also found necessary to have a 2 ft 3 in wide unit. However, the wider unit is more economical because obviously there is almost as much labour involved in making a 2 ft 3 in unit as in making one 4 ft 9 in—for example, in 1971 the cost of a 2 ft 3 in Type H cupboard unit is $60 compared with $88 for a 4 ft 9 in Type E cupboard unit, and a 2 ft 3 in Type J drawer unit is $75 as against $107 for a 4 ft 9 in Type D1 drawer and cupboard unit. This means that, especially in a large job, considerable savings can be effected by restricting the use of the smaller units to a minimum.

Details of the construction of a typical CSIRO unit are shown in Fig. 9. The ends, divisions and doors are framed and they have a honey-comb core, sheeted both sides with plywood; doors and drawers are fitted flush. Some manufacturers are now using particle board, veneered both sides. This involves less labour, but higher material cost; the price of the finished product therefore is about the same. For best quality work, framed construction is still recommended.

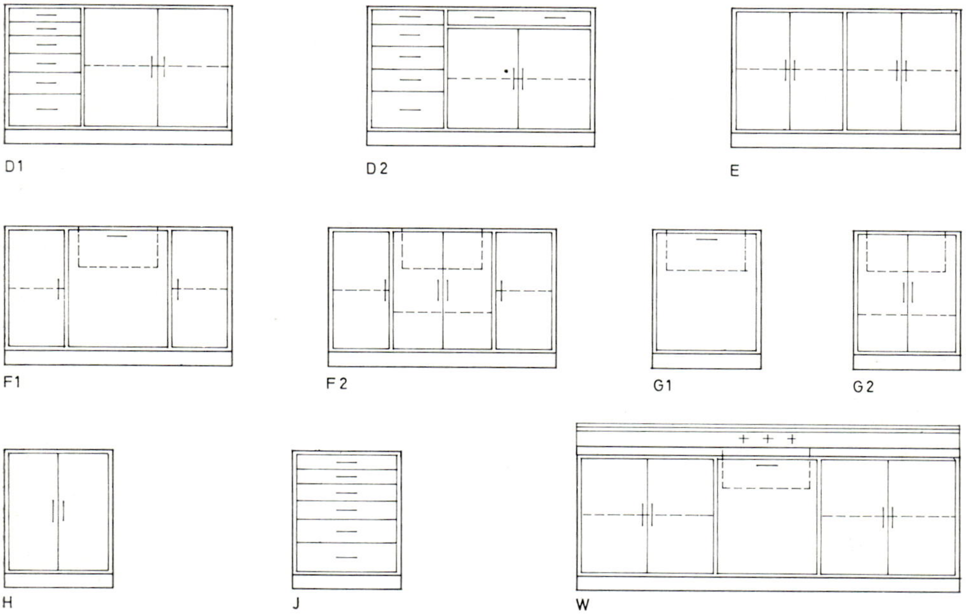

FIGURE 6 *CSIRO standard laboratory bench units.*

FIGURE 7 *View of a peninsular bench in a standard three-module laboratory; CSIRO Division of Land Research, Canberra, ACT. This layout is shown in Figure* 1.

By using a knee space with a combination of bench units, any specific length of bench can be obtained; furthermore, any on-site variation of length between partitions can easily be overcome by altering the width of the knee space. Also, a knee space makes it more comfortable to sit at a bench, it provides a convenient place for some equipment and it reduces the number of bench units, thus saving money. The width of a knee space should be no less than 2 ft, and no more than

FIGURE 8 *View of a wall bench; Biochemistry Laboratory, CSIRO Division of Plant Industry, Canberra, ACT.*

3 ft 6 in—about the maximum span of an unsupported bench top. If it is necessary for it to be wider, a front rail should be used up to 5 ft and a leg if it is wider still.

Benches are installed by the simplest method, *i.e.* the tops are supported by the drawer and cupboard units. When the service pipes have been installed and the service strip fixed above them, the units are placed in position (packed and scribed if necessary) and the tops are fitted. A slip joint is provided in the bench top at the junction of the service strip and the unit beneath; for wall bench installations where there are not many services, this joint is sometimes omitted and the one-piece top extends back to the wall. The service duct is 15 in. in peninsular benches and $7\frac{1}{2}$ in. in wall benches; access is obtained by removing screw-fixed panels at the backs of knee spaces and cupboard units.

In an effort to increase the flexibility of benches and accessibility of pipes, other methods of installation have been used; the trouble is that some of them give less and cost more. One method is to have the bench top supported on a steel frame and the units either fitted with casters or just pushed underneath; in a laboratory where the bench units must be mobile, or where very few units are required, this offers the best solution. Another idea is to have the bench units suspended from a steel frame and clear of the floor. Admittedly this gives a lighter and more modern appearance, but, if the space beneath the unit is sufficient to allow the floor to be properly cleaned, the reduction in the storage capacity of the unit is considerable—at little or no reduction in cost.

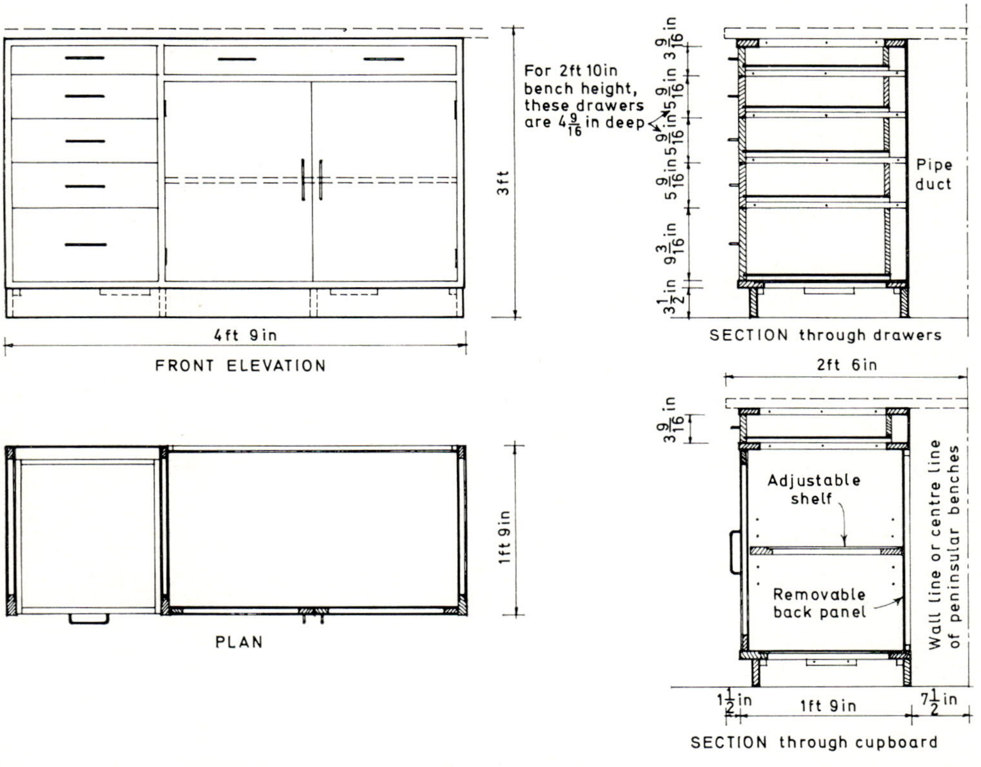

FIGURE 9 *Typical construction details for CSIRO standard bench units; the unit shown is Type D2.*

To quote one instance in an overseas laboratory—the benches were suspended 10 in above the floor, *i.e.* the height was reduced by nearly 20%. When the building was to be extended, the staff (very conscious of the loss of storage space) asked that the 10 in be reduced to 6 in; this, of course, made cleaning extremely difficult. The obvious query is why not reduce the height still further by standing the unit with its $3\frac{1}{2}$ in base on the floor? However, it is agreed that circumstances can alter cases. In a CSIRO laboratory recently built in the Northern Territory where there is a serious termite problem, a steel frame was used and the benches were suspended 6 in above the floor. Despite the floor-cleaning difficulty, this was considered the best solution because it made it easier to inspect for termite galleries.

An interesting recent development overseas is a cast epoxy bench top with integral backboard. It is installed by tilting it until the backboard fits into a rebate in the front of the raised service strip, and it is then lowered until it rests on the top of the bench unit—a simple and speedy operation. However, I understand that this type of bench top is available only in the United States of America at present, and it is expensive. The idea is good and it is quite possible that, with improved production, materials and finishes, it could become the almost perfect bench top.

Over the years it has been necessary to provide several variations of the standard bench units.

Mobile units

In some virology and bacteriology laboratories, where extreme cleanliness is imperative, the bench units should be mobile so that all wall and floor surfaces can be sterilised. Again, some scientists like to have a unit alongside them whilst they are working at the main bench, so it needs to be mobile.

Reversed units

There should be a nest of drawers beside each writing space. On one side of the bench, this is provided by a standard D1 unit; on the other side, it should be a D1 in which the nest of drawers is at the right instead of the left-hand end.

Drying cupboards

The two cupboards in an F1 unit are lined with asbestos board; each has a 250 W finned element and air inlet and outlet slots.

BENCH HEIGHT

For many years a bench height of 2 ft 10 in was accepted as being a satisfactory standard and there are still many scientists who prefer this height. However, 3 ft is becoming more general and is recommended. For some types of work, benches should be at table height, *i.e.* 2 ft 6 in.

BENCH TOPS

What is expected of a bench top? It should be solid, have a smooth, impervious, hard-wearing surface and it should withstand heat and be unaffected by chemicals; it should be hard enough not to scratch easily, but at the same time not be harsh on glassware.

Although there are many materials which very nearly meet all these requirements, to date there is not one which completely meets all of them. The widest range is available in the USA where the old reliable birch or maple with carbonised finish has been largely replaced by the following:

particle board with baked epoxy or plastic laminate finish;
cast epoxy sheet with integral coved back skirting and raised front edge;
Colorlith, a compressed asbestos cement sheet; it is available in several colours and in thicknesses from $\frac{1}{4}$ to $1\frac{1}{4}$ in;
Coloceran, a more recent development of Colorlith with a bonded ceramic-like inorganic surface; it is also available in several colours and in thicknesses from $\frac{1}{4}$ to $1\frac{1}{4}$ in;
Pyroceram, a glass ceramic sheet with a matt grey non-glare finish; it is available in sheets 8 ft by 4 ft and the thickness is 0·3 in.

Colorlith is slightly marked by some chemicals, but Coloceran and Pyroceram are completely resistant to all except hydrofluoric acid which causes slight etching. In this regard, they are better than timber, but it must be remembered that they are double or treble the price; another advantage is that they are available in colours and this fits in with present-day thinking. Points against them are that the surfaces are extremely hard and joints are required because the maximum sheet length is 8 ft.

In England, teak was commonly used, but in recent years substitutes such as iroko, teak face veneer on corestock and compressed asbestos cement sheet have become popular, mainly because they are cheaper. In Europe, ceramic tiles set in epoxy resin are common; again, these are completely resistant to all chemicals except hydrofluoric acid. Klingenberg tiles are available in several colours and in a range of special tiles for coves, return ends and pipes through the bench.

In Australia, timber is preferred for bench tops which will be subjected to the most severe chemical and physical treatment; compressed asbestos cement is also used, but to a lesser extent. Where conditions are less severe, tops sheeted with plastic laminate, vinyl, linoleum or plywood can be entirely satisfactory—for example, in a bacteriology laboratory, white plastic laminate would be considered by many as the best choice; for physical laboratories and instrument makers' benches, a vinyl or linoleum surface is excellent; stainless steel is the ideal material for radioisotope or oil laboratories. For isolated laboratories in hot dry climates, a solid timber top could be cracked or warped by the time it was delivered; compressed asbestos cement or a framed top sheeted with marine-quality plywood would be a better choice. Timber has the advantage that it is rigid, it extends the full length of the bench and there are no joints to contend with, and it provides a very pleasing working surface. As yet, no preparation has been marketed which will completely protect the surface from chemical attack and heat; but, with normal care and the use of bench mats, this need not be a real problem.

It will be seen, therefore, that the bench top material selected must be related to its specific use; of course, cost is a factor and sometimes geographic location.

Solid timber

The traditional timber for this purpose is teak, but in Australia it is expensive and the supply is poor. Queensland kauri is commonly used and is a very good substitute, but locally grown timbers such as ash in Victoria, jarrah in Western Australia, celery top pine in Tasmania and imported iroko (Africa) and kauri (Pacific Islands) have given satisfaction. Queensland kauri is a close-grained light-coloured timber with an even texture and low shrinkage, but it is a little softer than teak or ash.

It is important that the timber used in bench tops be first-quality—selected for colour if a clear finish is required. Accurate machining and jointing are essential and a high-quality polished surface is dependent on particularly fine sanding.

The tops are made up from boards generally not less than 6 in nor more than 10 in wide. Quarter-sawn (radial) timber is better than back-sawn (tangential) because the movement is less; one wide back-sawn board can cause warping and subsequent rejection of a top. With few exceptions, end joints should not be permitted. It is not necessary to use timber thicker than $1\frac{1}{2}$ in which gives a finished thickness of $1\frac{1}{4}$ in. Bench top projection is $1\frac{1}{2}$ in, with drip mould under.

Movement in timber tops

In order to avoid subsequent shrinkage, warping and cracking, it is essential that the timber be kiln-dried to a moisture content approximating that which it will attain in use; in Melbourne, this should be in the range of 9–12%. Timber will gain or lose moisture until it is in equilibrium with the relative humidity of the surrounding atmosphere. To illustrate this, moisture content readings of bench tops which have been in use for a number of years were taken at five locations in each of two buildings. In one which is air-conditioned, over a period of two years the moisture content varied between a maximum of 12½% and a minimum of 9¼%; in the other building which is centrally heated, during the same period the range was between 10¾ and 7½%. This moisture content variation will cause continual movement of the timber; surface coatings will retard it, but they will not prevent it. Movement per inch of width per 1% moisture content change for several timbers is shown below:

Timber	*Quarter-sawn radial (in)*	*Back-sawn tangential (in)*
Teak	0·001 0	0·002 2
Iroko	0·001 0	0·001 5
Queensland kauri	0·001 4	0·001 7
Victorian mountain ash	0·002 0	0·003 1
Jarrah	0·002 4	0·003 0

From the above it will be seen that, in a 2 ft 6 in wide bench top of quarter-sawn kauri, a change of 3% in the moisture content will cause a movement of approximately ⅛ in; with some timbers, this could be ¼ in. It is for this reason that solid timber bench tops must be fixed in such a way that it is possible for movement to take place, *i.e.* by using timber or metal buttons, or slotted holes. For all practical purposes, it can be assumed that solid timber does not move longitudinally.

The moisture content of particle board, when marketed, is generally within the range of 6–10%, and the movement between 0 (oven dry) and 25% (saturation) amounts to 7% in thickness and 0·7% in width and length. An increase of 1% in moisture content in a bench top 10 ft by 2 ft 6 in would result in an increase of 0·034 in. in length and 0·008 in. in width. In fact, an increase of 4% is not unusual; this would cause an expansion of slightly more than ⅛ in. in length and $\frac{1}{32}$ in. in width.

Movement in plywood is about 75% that of particle board. As 80% of this occurs in the moisture content range 0–8%, plywood can be regarded as having no movement under normal conditions, *i.e.* when the moisture content is above 8%.

Framed and sheeted

Select-quality timber suitable for bench tops is becoming increasingly difficult to obtain, and it is likely that framed and sheeted tops will be used more extensively in the future. Again, the recommended finished thickness is 1¼ in, *i.e.* ½ in ply on ¾ in frame; anything thicker is unnecessary and wasteful and anything less looks skimpy. Face materials are plastic laminate, vinyl, linoleum or plywood; end joints should be avoided, but as vinyl and linoleum are supplied in rolls, and plastic laminate sheets are available in lengths up to 12 ft, this problem arises only occasionally. If plywood is to be used, it should be marine-quality and be selected for sheet size, colour and appearance; if possible, the face veneer should be thicker than usual. A timber edge strip—tongued and glued—is recommended.

Finishes for timber bench tops

For many years, nearly all bench tops were finished with aniline black, and this is still used occasionally. Details of this preparation and its application are given in Appendix 3.

Nowadays there is a wide divergence of opinion regarding the most suitable colour for a bench top. Many scientists argue strongly that black is still the best because, with this background, a brown or purple stain is not conspicuous, and it is easier to read calibrations on glassware and to see small amounts of precipitate or any dust. Others favour a clear or coloured finish; they claim that black is funereal and that the possibility of stains does not really matter.

There are many epoxy resin, polyester and polyurethane preparations to give black, clear or coloured finishes. They all have very good (but not complete) resistance to acids, alkalis and organic solvents. A very smooth surface slightly increases resistance to chemicals; it is important, therefore, that the bench top should have a finely sanded surface with a buffed gloss finish.

In order to avoid an unequal moisture absorption which could cause warping, the underside of a timber bench top should be sealed, and two coats of a clear plastic are recommended; this also applies to plywood and particle board.

Bench mats

With reasonable care and the use of bench mats, a bench top will remain in excellent condition for many years. Suitable materials for mats are compressed asbestos cement, low-density polythene, coloured glass or Pyroceram.

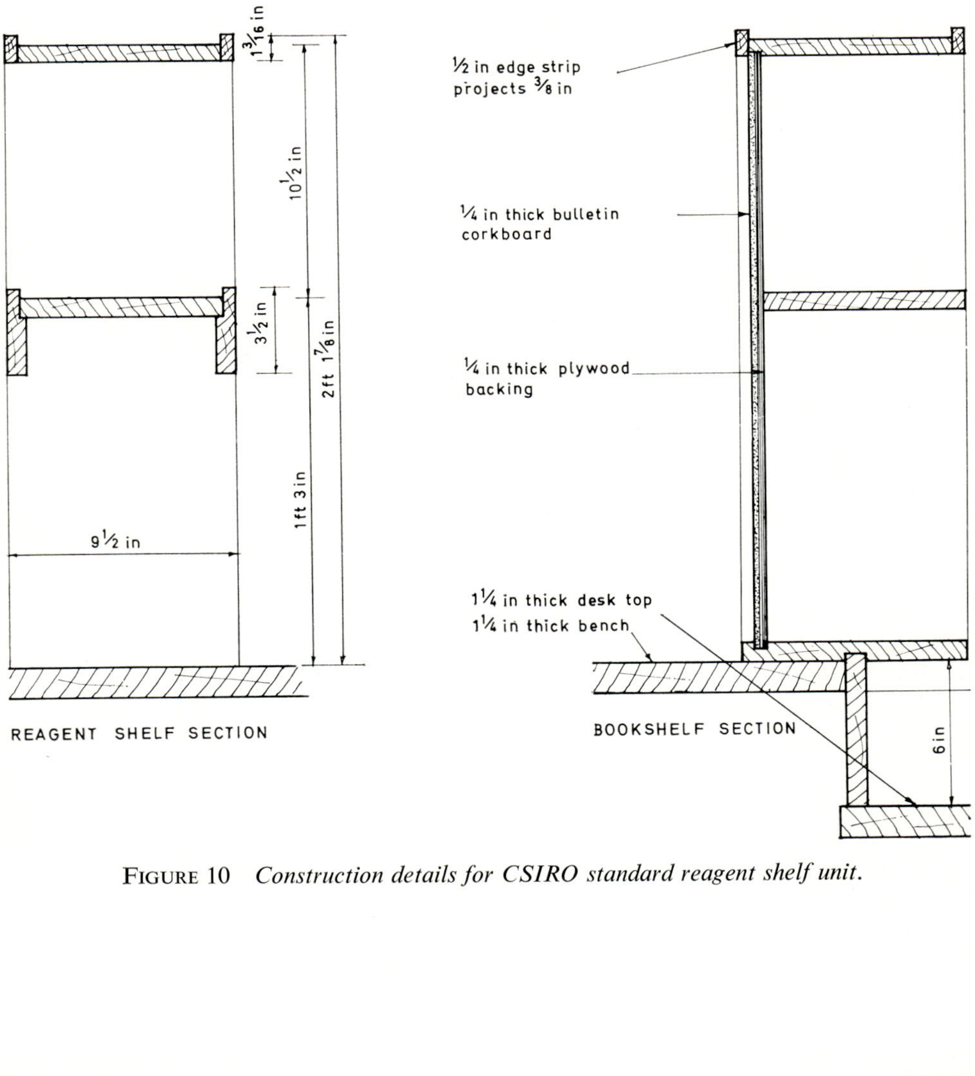

FIGURE 10 *Construction details for CSIRO standard reagent shelf unit.*

REAGENT SHELVES

In chemistry laboratories, most benches are fitted with reagent shelves. In some types of laboratories two shelves are required, but in others one shelf is adequate and it is a mistake to provide a second because it will only be cluttered with unnecessary bottles. The lower shelf and supports can be boxed in to accommodate pipes; this also provides a suitable mounting for service outlets (especially power) and keeps the bench top clear.

Details of the CSIRO standard fitting are shown in Fig. 10. The shelves have a raised edge to prevent bottles being pushed off accidentally. It is usual to protect the shelves; glass and stainless steel have been used, but low-density polythene sheet $\frac{1}{8}$ in thick, loose laid, provides a non-slip, chemical-resistant surface at less cost, and is recommended. Polythene saucers under individual bottles can also be used.

WRITING SPACES

Writing spaces should be provided for all laboratory assistants. A convenient place is at the window end of a peninsular bench, and the top should be 2 ft 6 in high with a plastic laminate or vinyl surface. The reagent shelves are extended to form bookshelves, and the back of these is faced with Krommenie bulletin-board cork for use on the opposite side of the bench. An alternative is to use 4 ft by 2 ft 6 in tables along one of the end walls; adjustable shelves can be fitted above.

SINKS

Materials used are stainless steel, vitreous china and PVC. It is a good idea to check the dimensions because quite often sinks are larger or deeper than really necessary; this means loss of bench space, a backache for somebody and additional cost.

Stainless steel

Nowadays stainless steel is used extensively for wash-up sink and drainer units, and for individual sinks and troughs for other than the most severe conditions. It provides a jointless surface which is not as hard as ceramic, and there is almost no limit to the design and size of units which can be fabricated to meet special requirements.

Individual sinks and troughs have a rim with fixing screws welded to the underside. This obviates a troublesome joint between bench top and sink, and the edges of the hole are protected—of particular value when the bench top is laminated.

In CSIRO there are sinks which have been in use for 20 years and are still in excellent condition. I have heard of only one case of failure—caused by weak hydrochloric which had been allowed to lie in the sink for several days; but, after all, this is not good housekeeping. Stains can generally be removed with steel wool, but it is important that an abrasive cleaning powder be used afterwards, otherwise rust marks will appear.

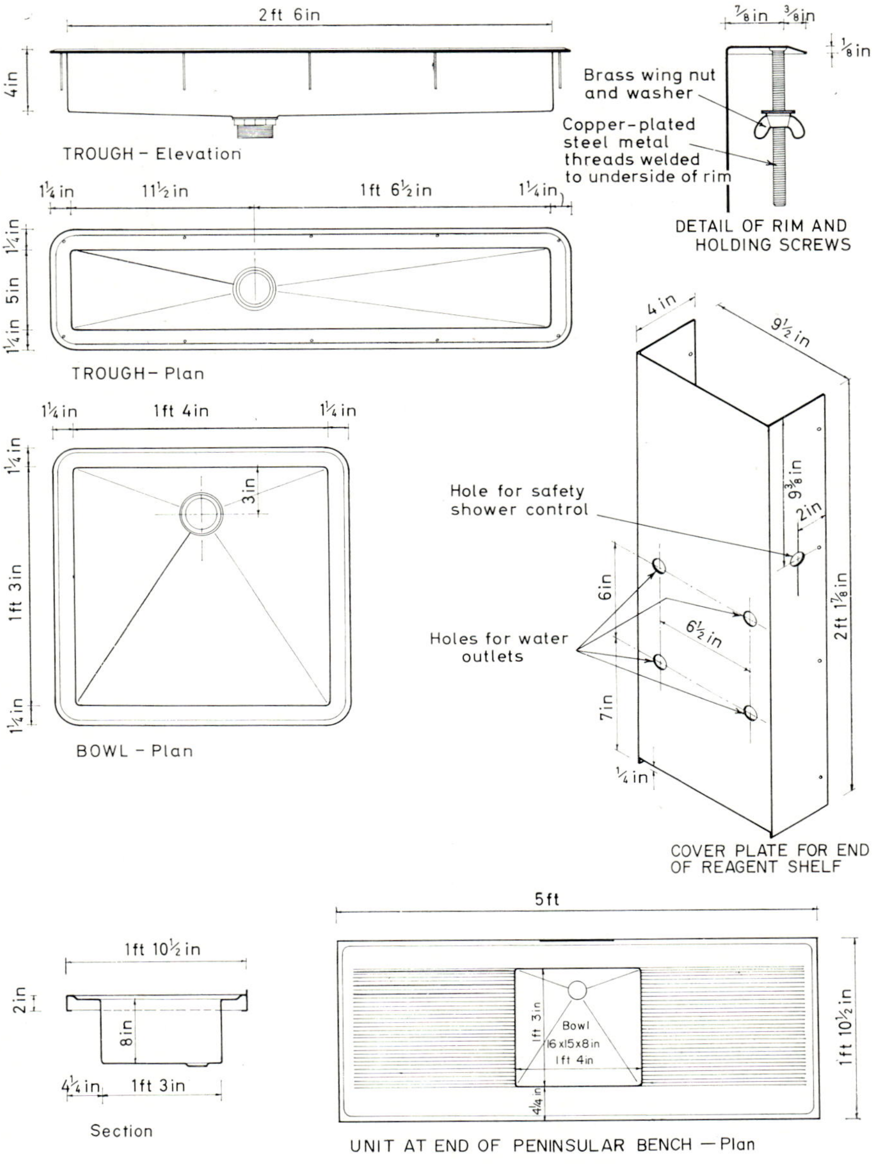

FIGURE 11 *CSIRO standard stainless steel laboratory sinks.*

Details of the CSIRO standard units are shown in Fig. 11. The grade of stainless steel used is No. 316–18/8 austinitic chrome nickel steel containing 2% molybdenum. It is manufactured in rolls in a number of widths; for minimum cost, consideration should be given to the width of the unit and the height of the back, etc., to avoid cutting to waste. The higher cost of molybdenum quality for laboratory use certainly is warranted; so also is the use of 18-gauge instead of 20-gauge. Satin finish is entirely satisfactory and therefore the extra cost of mirror finish is not justified.

The units have a backing of particle board glued to the stainless steel and this is protected on the underside with chlorinated rubber paint.

The following small points can be important:

For bacteriological work—where cracks and crevices must be avoided—the undersides of units should be sheeted with galvanised iron and the joints should be soldered.

In a bacteriology kitchen, a flat draining surface is preferable to the standard corrugated finish.

For animal post-mortems, the stainless steel unit should be supported by brackets on the wall rather than legs on the floor. Cleaning is also easier if the radius of internal angles is increased from the standard $\frac{1}{2}$ in to 1 in.

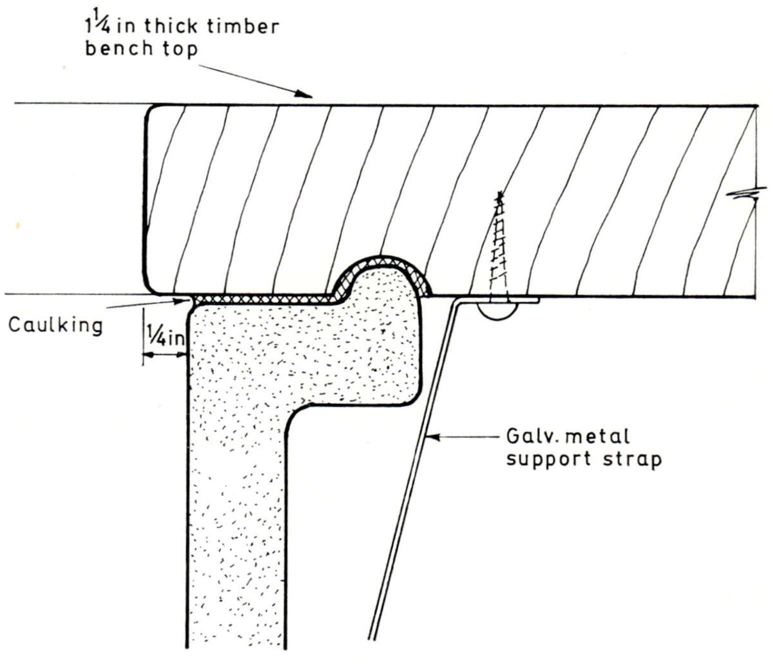

FIGURE 12 *Joint between bench top and vitreous china sink (R. Fowler Ltd Monash type).*

Vitreous china

The standard type has a straight flange and the outlet is bevelled to fit a Vulcathene or stainless steel plug and washer. The Monash type has a rebated flange which facilitates a water-tight joint to the underside of the bench top (*see* Fig. 12); it also has an integral outlet which is a decided advantage when severe chemicals are being used.

Vitreous china has high impact resistance, it is not subject to crazing and discolouration and it resists even hot concentrated acids and alkalis—provided, of course, that they are promptly washed down.

For trace element investigations where chromic acid is used to clean the glassware, the sink should be vitreous china and the drainer should be timber. There is a risk of the glassware being contaminated if stainless steel is used.

Plastics

Rigid polyvinyl chloride sheet can be fabricated into sink and drainer units, tanks and trays; the sheet is generally $\frac{3}{16}$ in thick and the joints are welded. Again, particle board backing should be glued to the underside of the drainer and also to the sink—in case hot water is used; the backing should be finished with chlorinated rubber paint.

TROUGHS AND DRIP CUPS

In student laboratories, full-length runnels of stainless steel, PVC or polythene are used quite often, but in research laboratories troughs or drip cups are preferred because they occupy less bench space.

The standard CSIRO trough 30 in by 5 in by 4 in (*see* Fig. 11) is used with a four-tap fitting over (*see* Figs. 1 and 7). Individual 7 in by 4 in moulded Vulcathene drip cups are available, but many laboratory workers consider them too small and prefer a 12 in by 6 in or 8 in by 6 in stainless steel trough.

TAPS

In the early 1940s, wartime difficulties of supply made it necessary for CSIRO to design its own laboratory taps and arrange their manufacture and these are still being used. In recent years, a comprehensive range of well-designed imported taps has become readily available commercially and these are now being used quite extensively.

The CSIRO taps are brass—$\frac{1}{2}$ in jumper type; the tubing is $\frac{13}{16}$ in by 13-gauge brass and the joints are brazed. Hexagons are provided on all fittings to facilitate fixing; tapered nozzles and adapters are available for screwed connections. Most of these fittings are chrome plated, but grey epoxy is becoming increasingly popular and is the finish used on swan-necks, etc. in fume cupboards.

The imported Marklab valves are brass—$\frac{3}{8}$ in with smooth easy-clean surfaces. The tube is brass with a minimum thickness of 10-gauge; swan-necks are made from 16-gauge tube and they are either fixed or swivel. All types of bench and wall fittings are supplied with integral screwed tails and backnuts with locating lugs which eliminate screw-holes in the flanges. The standard finish for all fittings is black polythene and they have handwheels with lettered coloured inserts. Recently the jumper valve has been largely superseded by a diaphragm valve of improved design and performance; the rubber diaphragm, when closed, is in the relaxed position and when open it is compressed. These valves are glandless and therefore friction is greatly reduced. They are designed to control hot and cold water, gas, compressed air, vacuum and high-pressure gases; for distilled water, special fittings are available, including a swan-neck in which all the contact surfaces are plastics. Needle valves suitable for up to 100 lb/in^2 are available; they have stainless steel floating-tip spindles which self-align into a brass seat.

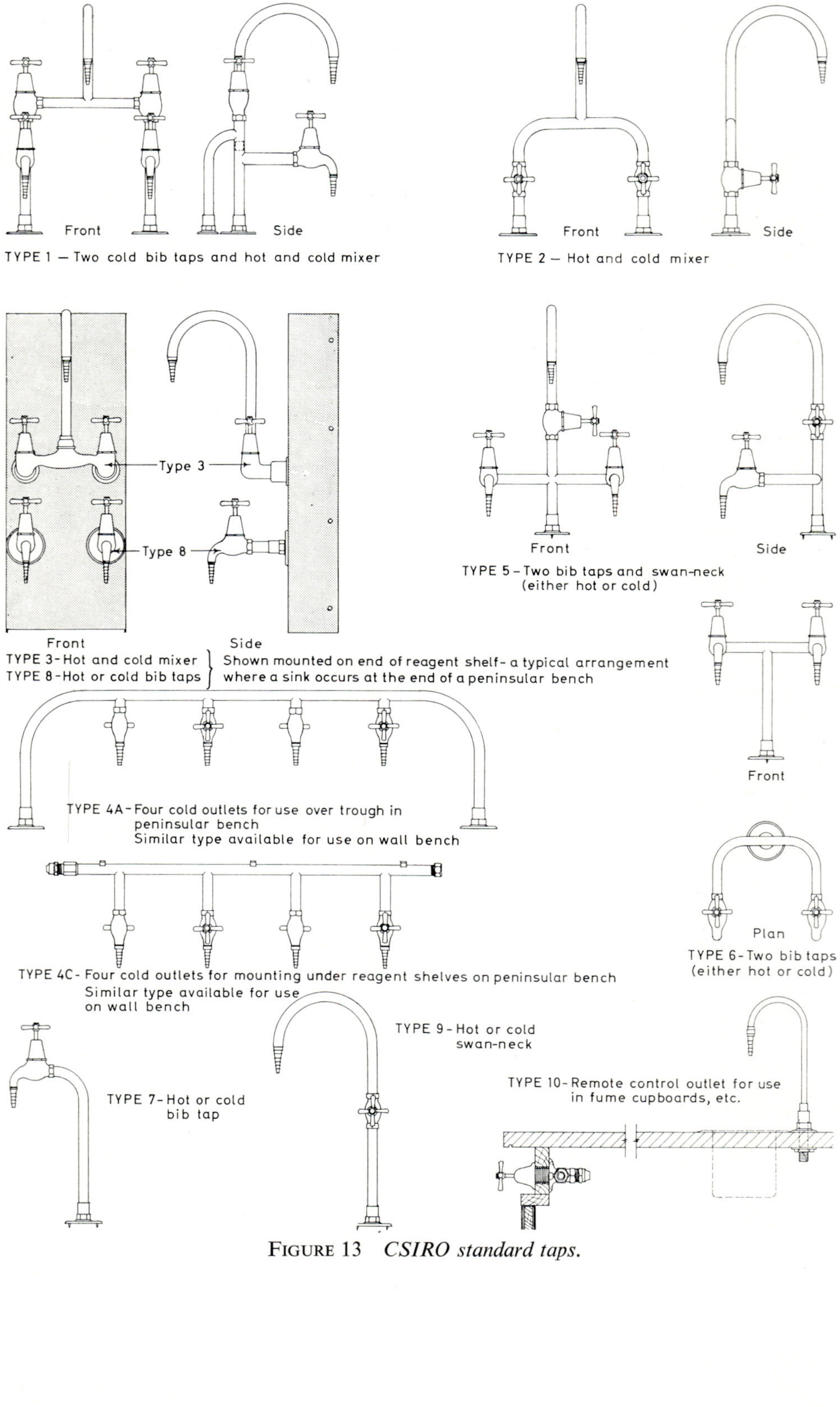

FIGURE 13 *CSIRO standard taps.*

DRAINING RACKS

Peg boards are used, either made of timber or, occasionally, stainless steel; in both cases, the ends of the pegs are fitted with plastic caps. Nowadays polythene-coated wire racks are gaining popularity; they are less expensive and can be made up to any size (*see* Figs. 7 and 8).

The functions of a draining rack and a drying cupboard have been combined in a unit marketed in the United States of America; it is fitted with a heater and fan, and air at 160°F is blown through the pegs which are polypropylene tubes.

FUME CUPBOARDS

The function of a fume cupboard is to protect laboratory personnel from dangerous fumes and to contain any minor fires or explosions which might occur. It must be designed to collect and disperse the fumes without risk of them re-entering the laboratory through windows or fresh-air supply ducts. The quantity of air from the laboratory required to purge the cupboard must be a minimum, and air turbulence within the cupboard must also be a minimum. All internal surfaces must be chemical-resistant.

Nowadays more and larger fume cupboards are being used and it is vitally important that location, design and the materials of construction be given careful consideration.

Location

For maximum economy and efficiency, a fume cupboard should be located so that the exhaust duct can be vertical; bends and horizontal runs should be kept to an absolute minimum. It should be in a draught-free position because excessive air movement across the face of a cupboard will prevent it functioning properly. The velocity of the air being exhausted through a cupboard is quite often less than that of air currents entering the room through an open door or window or through air inlet registers; it can even be less than air disturbance caused by a person walking in front of the cupboard. Also, a fume cupboard should be in a position where, in the event of fire, access to the door is not obstructed; of course, if there is an alternative exit, this point does not arise.

In buildings where there are heated-air or air-conditioning systems, there is an advantage in having some of the fume cupboards grouped in rooms between laboratories. If this is done, each room has its own supply of partially heated air.

Design

The satisfactory performance of a fume cupboard requires that contaminated air does not reach the breathing zone of the operator. To achieve this, tests indicate that the velocity across the face area of the fume cupboard opening should be uniform and not less than 50 ft/min; in practice, 60 ft/min is generally adopted as a safe working minimum and this is increased to more than 300 ft/min when the sash is lowered to give a face opening 6 in high. It can be seen, therefore, that with higher overall face velocities, the incoming air can cause problems; there can be excessive turbulence, gas flames can be blown out and, in fact, the air can rebound out of the cupboard.

In order to obtain an even distribution of air across the fume cupboard, baffles are fitted at the back and at the top, thus forming two (or preferably three) full-length slots. Heavy fumes are exhausted through the bottom slot and light fumes and heated air through the top back slot; the top front slot picks up any fumes which have not been drawn into the back slot and have rolled forward.

Types of fume cupboards

Open front

This is the simplest and least costly cupboard and it is entirely satisfactory where conditions are not severe.

Sash

Details of the CSIRO standard cupboard are shown in Fig. 14. It has baffles at the back and top and these provide three slots the full length of the cupboard. The bottom slot is $1\frac{3}{4}$ in wide and 7 in above the bench top; the two top slots are $1\frac{1}{4}$ in wide. One 40 W 4 ft fluorescent tube is fitted into a vented box which is either painted or lined with white PVC; it is protected from acid fumes by a $\frac{1}{4}$ in thick Perspex panel with neoprene gasket. A plenum is provided and the outlet to the duct is central at the top. The mullions are fitted with removable covers, providing a place for cables to power outlets and switches for the exhaust fan and light; it also accommodates the occasional pipe to service outlets.

Bypass

The bypass cupboard has been developed to overcome the problems which can occur if there is a high-velocity jet of air when the sash is in the low position. In this type of cupboard, an inlet grille is provided above the sash, as shown in Fig. 15. When the sash is fully raised it blocks the inlet grille and all the air enters through the sash opening. The quantity of air entering through the grille increases as the sash is lowered, but the combined volume through it and the sash opening is constant, *i.e.* regardless of the position of the sash, the quantity of air exhausted from the laboratory does not vary and this avoids complications with ventilating or air-conditioning systems.

For air-conditioned laboratories

Air-conditioning—particularly cooling—is an expensive business, and to exhaust conditioned air through fume cupboards seems to be very wasteful. Obviously this is a major consideration in the design of air-conditioning systems for laboratories and the subject is discussed in some detail in Chapter 5.

In order to reduce the quantities of conditioned air exhausted, much attention has been given to the design of fume cupboards with auxiliary air supply. A separate fan and duct system can supply outside air (partially heated when necessary) up to 60% of the total quantity required. The difficulty here is to introduce the auxiliary air into the cupboard by a method that will maintain an adequate uniform velocity over the face opening and yet not cause turbulence within the cupboard.

In Australia, Conditionaire Engineering Pty. Ltd market a fume cupboard in which the auxiliary-air inlet is inside the cupboard; in order to reduce the velocity, the air passes through gauze. In England, some fume cupboards have been manufactured to a design described in an article 'Sashless Fume Cupboards' by M. J. Sanders, Safety Officer at the Ministry of Technology's

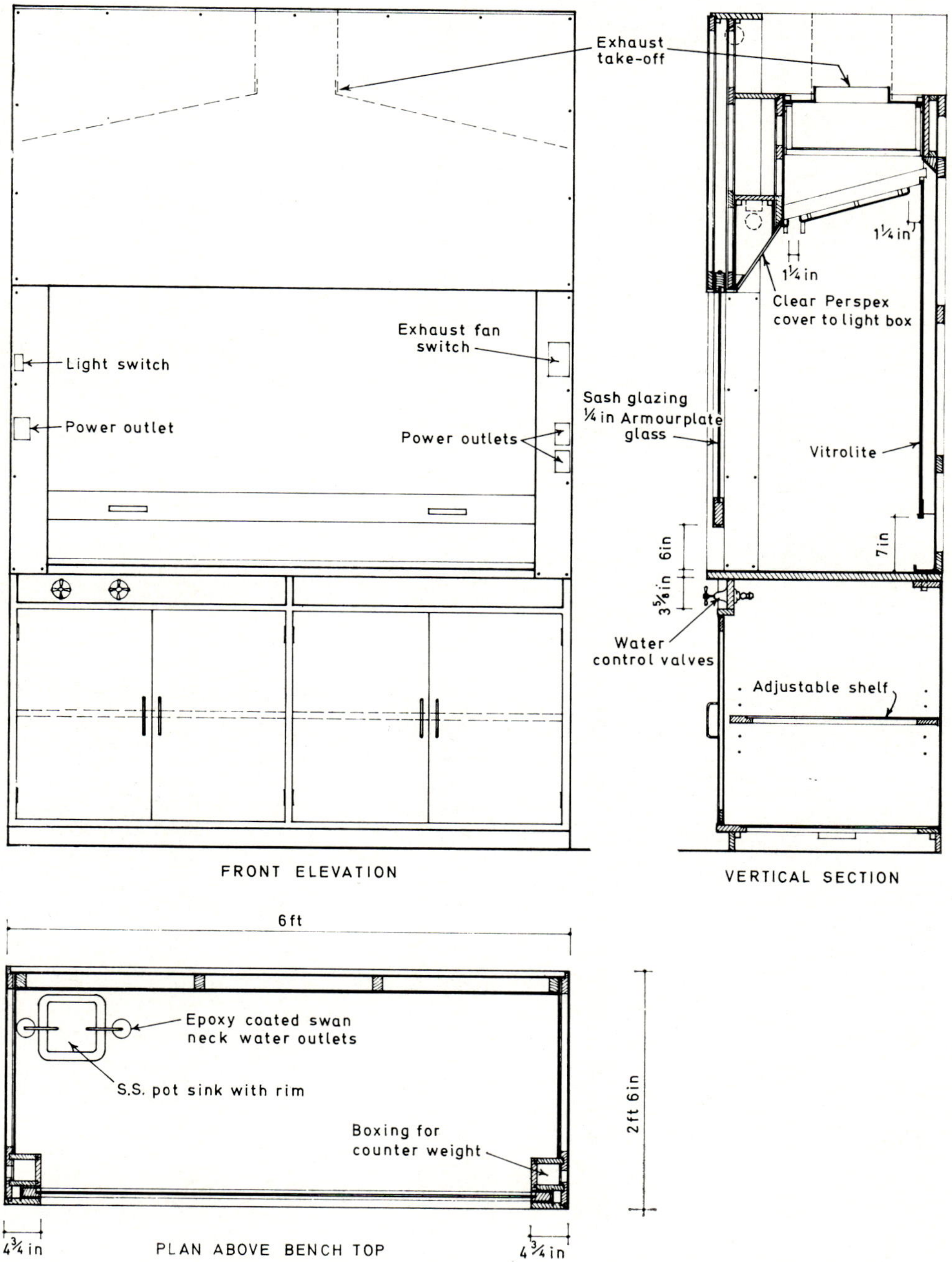

FIGURE 14 *CSIRO standard fume cupboard with sash.*

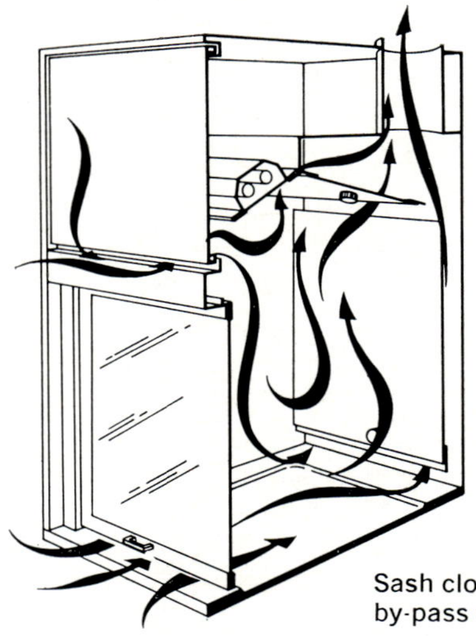

FIGURE 15 *Bypass fume cupboard, marketed by Fisher Scientific Co., USA.*

Warren Spring Laboratory. In both designs, the auxiliary air is intended to form a curtain across the face opening. In the United States of America, most of the auxiliary-air-supply fume cupboards are designed so that the air enters through a louvred panel which projects from the fume cupboard above the face opening, as shown in Fig. 16.

Many scientists say that they are satisfied with the performance of these types of cupboards; in fact, there is some doubt that the flow pattern of the auxiliary air is as claimed. This matter

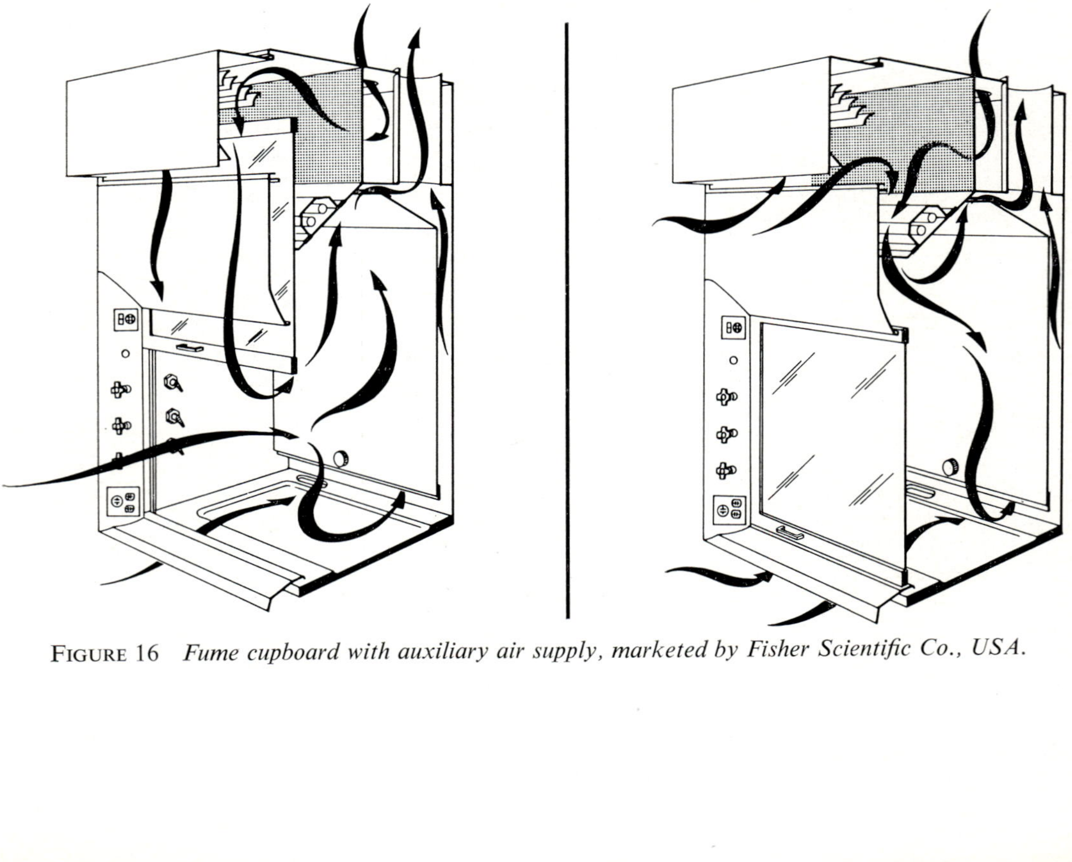

FIGURE 16 *Fume cupboard with auxiliary air supply, marketed by Fisher Scientific Co., USA.*

has been examined and several reports written; in one, 'Evaluation of Laboratory Fume Hoods' by Schulte, Hyatt, Jordan and Mitchell (*American Industrial Hygiene Association Quarterly*, September 1954), the authors state:

> 'It is this detrimental effect of cross drafts which nullifies attempts to provide an independent source of unconditioned or untempered air to a hood. Tests were made on one hood where air was brought in along the sides and top of the hood face to supply about one half of the total amount of air exhausted by the hood. Correspondence with other engineers and observations made in other laboratories have convinced us that the independent air supply is not a solution to the problem of saving heated or cooled air.'

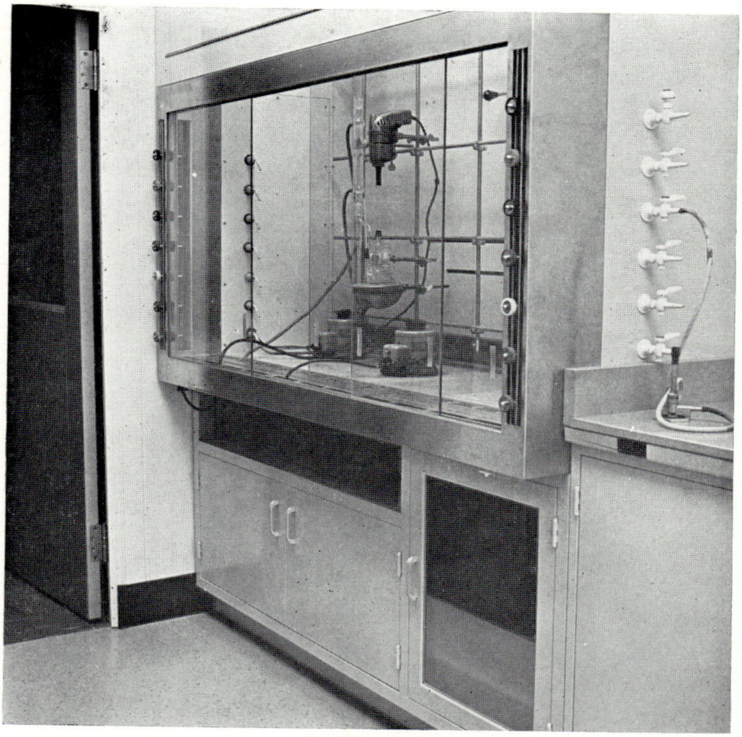

FIGURE 17 *Fume cupboard; Monsanto Chemical Co., Missouri, USA.*

There is another way to solve the problem, *i.e.* to design the fume cupboard so that, whilst the air exhausted is conditioned, the quantity is reduced to a minimum by reducing the area of the face opening. This matter was investigated very thoroughly by the Monsanto Chemical Co. in connection with the design of their Research Centre at Creve Coeur, Missouri. A 6 ft wide fume cupboard was to be provided for each of the four men to be accommodated in each of the 80 two-module 22 by 30 ft laboratories. I quote the following extracts from the report 'Fume Hoods—Safety vs. Costs' by Walls and Metzner:

> 'The hood design must satisfy the size requirements for conducting experiments in all branches of chemical research except radiological.

'It must provide a safe face velocity and use no more than 550 ft^3/min of room air, *i.e.* four hoods require 2200 ft^3/min. An excess of this quantity would result in high discomfort to the occupants. It must be easily balanced as a part of the total air-conditioning system. No air from the laboratory rooms is to be recirculated in order to reduce the danger of experiment contamination.

'The rule-of-thumb correlation between fume hood exhaust and air-conditioning refrigeration is that each 200 ft^3/min exhausted equals one ton of refrigeration which represents \$1000 in capital investment. Thus a conventional 6 ft fume hood having a face velocity of 100 ft/min and an area of 16·5 ft^2 would require an air-conditioning system costing \$8200, to say nothing of the operating cost.

'The conventional fume hood was obviously not acceptable. A very high volume of air (up to 1600 ft^3/min) is required to produce safe face velocity when the hood is open far enough to permit freedom of work. The auxiliary-air hood with internally supplied air was dismissed because it accomplishes absolutely nothing toward reducing the air required from the room to assure a safe face velocity. In fact, our experience with this type of hood was that it tended to "bounce" air out of the hood. The auxiliary-air hood, with supplementary air supplied just outside and pulled through the face of the hood, is a satisfactory choice from the safety standpoint. Again, large quantities of air must be handled. Both of the auxiliary-air hoods must be provided with a bypass to take air from the room when the sash is closed, or the system will be hard to balance. Such a system is also expensive because it requires additional ductwork.

'A prototype was installed and tested in a full size laboratory mock-up. Details of the final design are shown in the photograph (*reproduced in this book as Fig.* 17), and incorporated the following features:

(i) A three-section horizontal sliding sash which yields a face velocity of 120 ft/min in the one third open position and 55 ft/min in the two thirds open position at a total air discharge of 550 ft/min.

(ii) Above the sash is a hinged blow-out panel held in place by magnets. This could relieve pressure within the hood in the event of an accident and reduce the chances of the Saflex safety glass blowing completely out of the frame.

(iii) In order to maintain a balance in the air-conditioning system, the air flow is through the hood at all times. It enters through simple baffles at the sides and along the bottom when the sashes are closed, and through these and the face opening when it is partially open.'

Perchloric acid

Because of the extremely dangerous properties of perchloric acid, all laboratory work involving its use should be done in specially designed fume cupboards which conform to the following:

All internal surfaces should be incombustible and non-porous—for example, polyvinyl chloride (PVC), stainless steel or compressed asbestos cement with epoxy paint finish.

To prevent a build-up in concentration of perchloric condensate, PVC or stainless steel water sprays should be fitted behind the baffle and in the duct. It is important that the position and design of the sprays be such that they wash the entire surface. The water is collected in a trough which extends full length along the back of the cupboard and in one or more drain

outlets in the duct. The sprays can be operating whilst the cupboard is in use or they can be turned on at the end of each digestion. Water falling from the bottom of the back baffle can cause trouble by splashing; in one design, this has been solved by using a strip of polypropylene fibre. The baffles should be removable to simplify regular and thorough maintenance.
Any unsealed crevice is a hazard. All internal surfaces—in both the cupboard and the duct—should be streamlined to avoid ledges where condensate can lodge.
Timber should not be used; even when coated with epoxy or chlorinated rubber, it is subject to attack at joints. Litharge and glycerine should not be used as a cement for bedding plugs and washers in sinks; it absorbs perchloric acid and overheating or percussion can cause it to explode.
Ducts should be stainless steel (avoid the use of rivets) or asbestos cement with chlorinated rubber paint finish.
Each cupboard should have a separate exhaust.

Spot ventilation

Spot ventilation in the laboratory provides a flexible means of exhausting contaminants and, as this is done near their source of origin, relatively small quantities of air are required. Fumes can be exhausted from a small enclosure on the bench or from washing-up sinks and drying cupboards. A mobile unit can be used, as shown in Fig. 18; this type of installation has been simplified by the use of plastic enclosures and flexible PVC ducting.

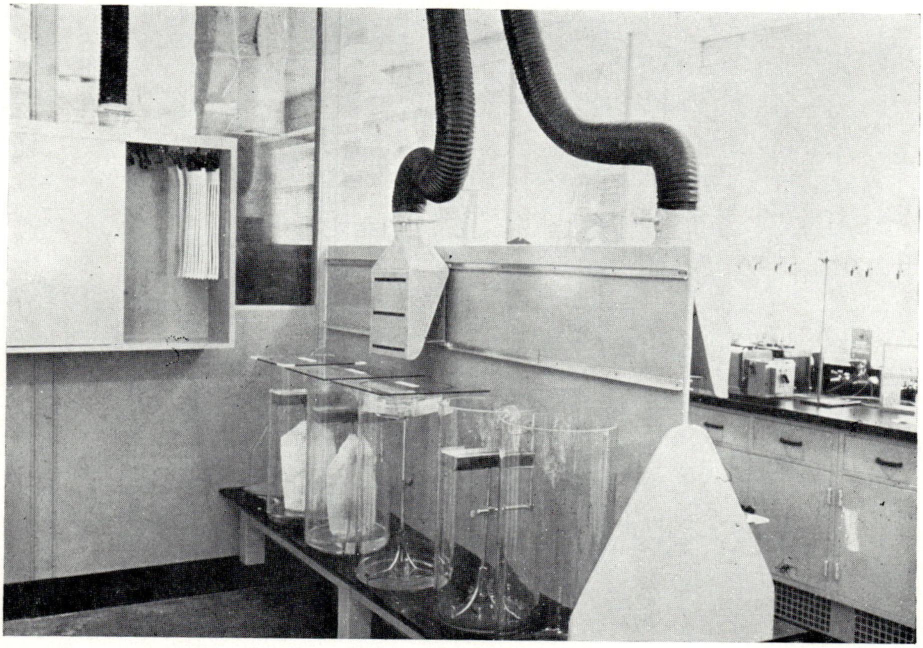

FIGURE 18 *Showing slotted exhaust units for spot ventilation; Upjohn Laboratory, Michigan, USA. The 5 in diameter flexible ducts each exhaust 300 ft^3/min. The units are fitted with shut-off dampers, non-sparking wheels, and sliding track. Solvent vapours are exhausted from the drying cabinet at the left at a rate of 100 ft^3/min.*

Dimensions

For many years laboratory workers were satisfied with a fume cupboard 4 ft wide, but now there is an increasing preference for cupboards 6–8 ft wide. For practical reasons such as air distribution, duct size and width of sash, 8 ft represents about a maximum. As in the case of bench units, an increase in length means less cost per lineal foot; the installed cost of a 6 ft cupboard, complete with services, is very little more than for a 4 ft cupboard.

A clear height of 4 ft is generally adequate and is recommended, but this should be checked because, for some work, 3 ft is quite sufficient. Then again, heights of 5 ft or more are sometimes required, and these can be obtained either by having a portion of the bench removable or by having a walk-in unit. Most cupboards are 2 ft 6 in deep, but current designs in the United States of America are 3 ft which includes a splayed surround 6 in deep; the extra depth probably improves the air flow into the cupboard.

The CSIRO standard design provides cupboards under the bench top, but variations include acid store (lined with asbestos or PVC, and fitted with a PVC tray), chromatography cabinets or an open shelf for electrical equipment.

Construction

In the CSIRO standard fume cupboard, the frame and sash are timber. The bench top is timber, compressed asbestos cement, stainless steel or acid-proof tiles set in epoxy resin; if tiles are used, the size of the bench top and the sink should be an exact multiple of the tile plus joint, to avoid cutting.

The internal lining is rigid PVC, $\frac{1}{8}$ and $\frac{3}{16}$ in thick—white where it is visible and grey (which costs less) behind baffles. This material is not attacked by chemical fumes and its smooth polished surface has a pleasing appearance; with the careful application of heat, it can be bent to form condensation gutters, etc.

The back baffle is coloured glass—generally white, but black has been used because it is a better background for colour assessment of fumes. To reduce the risk of cracking, it is fitted in panels 2 ft wide which are butt-jointed and supported by clips rather than screws. Timber or stainless steel is used to support the back and top baffles; timber should have two coats of epoxy before fitting. The top baffle is white PVC with edges bent up; this gives added strength and, along the lower edge, serves as a condensation gutter.

Of course, all materials in a fume cupboard should be incombustible, and therefore PVC cannot be regarded as being completely satisfactory. It will distort if subjected to temperatures of about 150°F, and at higher temperatures it will burn—giving off hydrochloric acid fumes; however, it is self-extinguishing. Even when several hot-plates are being used, the large quantities of air passing through the cupboard make it most unlikely that the temperature would rise sufficiently to distort the PVC; if this does happen, it is probably due to carelessness. To avoid this possibility, some people prefer to have the cupboard (including the back and top baffles) lined with $\frac{1}{4}$ in compressed asbestos cement, painted with clear or white epoxy. For some types of work, the lining should be stainless steel which gives flush surfaces and can be coved at internal angles.

Sashes

When a fume cupboard is being used for light-duty work, it is serving the function of an exhaust hood and a sash (quite an expensive item) is not really necessary. However, for most laboratory work, a fume cupboard should be fitted with a sash because it improves the performance and provides a safety barrier for the operator. Sashes are sliding—horizontal or vertical, or a combination of both. In air-conditioned laboratories, sashes which slide horizontally have the advantage that they reduce the face area of the cupboard (*see* p. 54).

The CSIRO standard fume cupboard has a sash which slides vertically. It has a timber frame and the glass is bedded in an adhesive glazing compound in a $\frac{5}{8}$ in rebate on the inside of the frame. The weights are lead, the wire stainless steel and the pulleys nylon. In order to work in the fume cupboard without raising the sash, small horizontally sliding panels are sometimes provided along the bottom of the sash; alternatively, there can be glove holes. The glass is $\frac{1}{4}$ in thick Armourplate, and some properties of this glass relative to wired glass and laminated glass are set out in Table 1.

TABLE 1. Some properties of toughened, wired and laminated glass

Type of glass	*Resistance to thermal shock*	*Mechanical strength*	*Safety properties in event of explosion*
Toughened—*e.g.* Armourplate	Withstands up to approx. 480°F	4–5 times as strong as polished plate of the same thickness	The glass will shatter, but the particles will be small with relatively blunt edges
Wired	Less than for polished plate, for which the maximum is 120°F	Less than the same thickness without wire	The glass and possibly the wire will break, and sharp-edged particles could be projected at high speed
Laminated—*e.g.* TripleX	Withstands up to approx. 120°F	Same as polished plate of which it is comprised; thus $\frac{1}{4}$ in TripleX is only as strong as one of the $\frac{1}{8}$ in laminations	The glass will fracture and the plastic interlayer may rupture, but the particles will remain bonded to it

If Armourplate is used, it is cut to size and any required holes are drilled before the toughening process. Holes for handles and sash cords reduce the strength of the glass and therefore should be avoided; if they are really necessary, they should be fitted with nylon or red-fibre bushes because it is not good practice to have direct contact between glass and metal. During the manufacture and installation of the fume cupboard, the Armourplate should be handled with extreme care because a knock can produce a stress condition which, although impossible to detect at the time, can cause the glass to shatter if it receives another knock or is exposed to a sudden temperature change.

Testing and maintenance

Every fume cupboard should be thoroughly tested before it is accepted. A velometer is used to obtain the air velocities at, say, four positions across the bottom, the centre and the top of the face opening; readings should also be taken along the length when the sash is down to heights of 12 in and 6 in above the bench. The easiest method of testing is with an electrically operated portable insecticide fogger which generates large quantities of dense fog, although the high velocity discharge can be a disadvantage for some tests; in this case, smoke should be used. In Australia, the source of smoke is usually improvised, but smoke candles can be obtained—for example, E. Vernon Hill & Co., Wisconsin, USA, market two sizes; one is a half-minute candle which produces 4000 ft^3 of a light-grey, odourless and non-toxic smoke, and the other is a 3-min candle which gives 40 000 ft^3.

Even if it remains a pious hope, it would be good to think that all fume cupboards—including the ducts, fans and motors—were cleaned and checked once a year.

BALANCE AND INSTRUMENT BENCHES

The location of balances and sensitive equipment is most important, and the following points should be considered:

Under no circumstances should it be possible for sun to shine directly on to instruments.

Excessive air movement must be avoided; instruments should not be exposed to draughts from doors or windows, or to air currents from central-heating radiators. If the room is air-conditioned, care must be taken with the design, particularly the type and position of vents.

Good natural and artificial lighting is essential so that the operator is not bothered by either glare or shadow. Facing a window is not suitable; even a south-facing window generally is unsatisfactory because of glare. Quite often a magnifying attachment is fitted to a balance—to prevent glare on the scale, rather than to magnify it.

Instruments should not be placed in a constant traffic way.

In order to obtain the most rigid base, instruments should be set up over beams rather than slabs and close to columns rather than mid-span.

Balances and instruments in frequent use must be strategically placed. Current practice is for weighing to be done within the laboratory rather than in a common balance room along the corridor. Quite often semi-microbalances (precision $\pm 0{\cdot}01$ mg) are provided in the laboratory; in many, satisfactory readings can be obtained from a macrobalance ($\pm 0{\cdot}05$ mg), even if it is supported on a timber bench top.

If trouble with vibration is suspected, a simple test is to place a beaker of mercury on the floor and observe the surface; instruments for measuring the amplitude are also available. Various methods of exciting the building structure should be tried—for example, a person running heavily along a corridor or slamming a cold room door.

A simple bench with a heavy top will meet most requirements if, in the planning and design of the building, adequate precautions have been taken to deal with vibration (*see* p. 25). In basement or ground-floor rooms, the recommended bench is a 2 in slate or $2\frac{1}{2}$ in thick reinforced concrete or terrazzo slab on brick or concrete piers; the latter should be independent of the structure and the floor. For rooms above ground level, a similar top can be used on a steel frame, but there must be vibration-isolating material between them. Details of the CSIRO standard balance and instrument bench are shown in Fig. 19.

With the modern balance, weighing is generally a standing rather than a sitting operation; an increased bench height therefore provides more convenient reading. Assuming that the optical scale is 1 ft 2 in above the slab, a total height of 4 ft 6 in can be obtained by increasing the bench height from the standard 3 ft to 3 ft 4 in.

An instrument should maintain its level so that it is not necessary to make frequent adjustments. As desiccators, etc. have to be placed on the bench top and readings recorded, it is sometimes desirable to have the writing surface independent of the balance support.

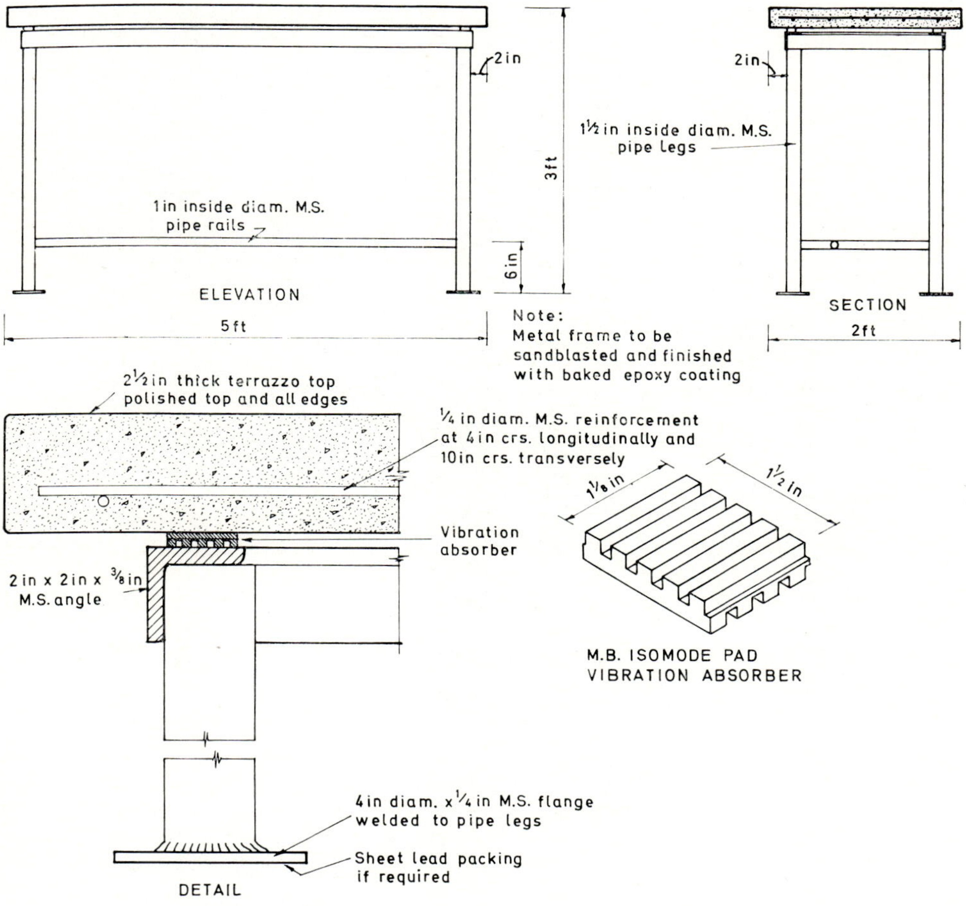

FIGURE 19 *CSIRO standard balance and instrument bench.*

The most commonly used vibration-isolating material is ribbed neoprene; the ribs on one side are at right angles to the ribs on the other, as shown in Fig. 19. The maximum efficiency is obtained when it is loaded to 50 lb/in^2 and the area of each pad is calculated accordingly; at this loading, the thickness of $\frac{5}{16}$ in is compressed to $\frac{1}{4}$ in.

Air bags have been used quite effectively to isolate vibration. At the David Rivett Laboratory (*see* p. 104), an electron microscope was installed on a suspended floor; photographs of up to 200 000 magnifications and exposures of up to several seconds were required. The equipment (which weighs approximately half a ton) is mounted on eight air bags, a $\frac{1}{2}$ in thick steel base plate and eight 6 in by 6 in Isomode neoprene pads (*see* Fig. 20). The air bags are butyl rubber of the type commonly used in motor car heavy-load air springs; they are approximately 4 in. in diameter by 7 in long, with a wall thickness of $\frac{5}{16}$ in. No additional damping is required, but

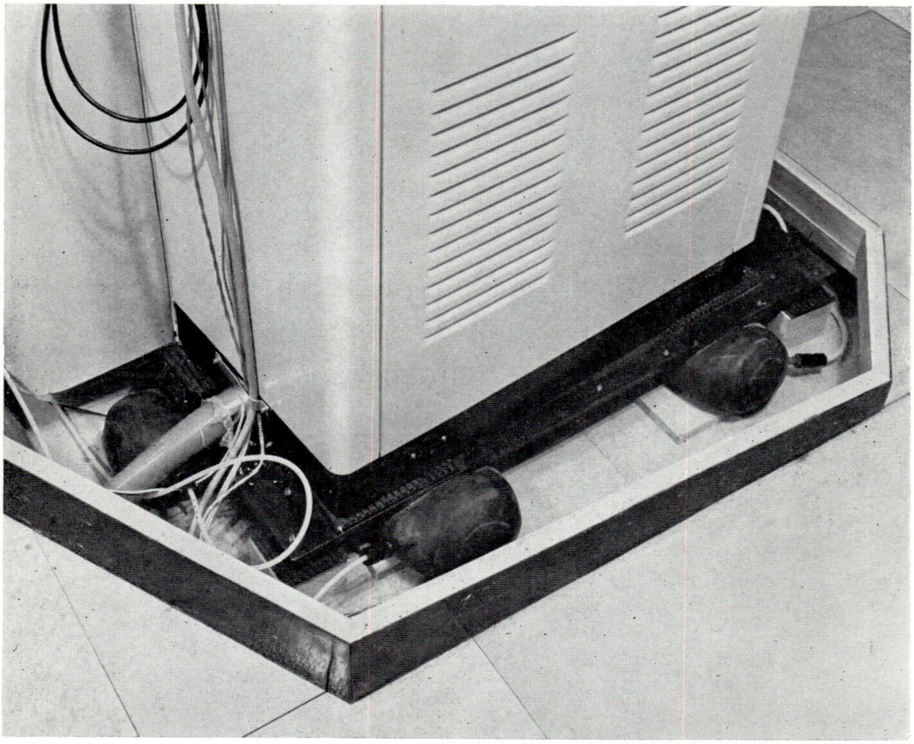

FIGURE 20 *Anti-vibration of electron microscope, using air bags and Isomode pads; CSIRO David Rivett Laboratory, Clayton, Vic.*

blocks with about $\frac{1}{16}$ in clearance have been placed at the four corners to restrict excessive movement. A troublesome vibration was transmitted through the high-voltage lead to the top of the machine; this was very simply and effectively absorbed by supporting the lead on two spiral springs fitted to the wall.

MICROSCOPE CUPBOARDS

The modern microscope with its numerous attachments is not readily stored in an under-bench cupboard; most scientists prefer to have it housed in a cupboard on the bench. Details of the CSIRO standard unit are shown in Fig. 21; the depth of the instrument should be checked because sometimes it is necessary to increase the cupboard from the standard 2 ft 3 in to 2 ft 6 in. The internal finish is a matter of personal preference; some like matt black, others white and others again prefer the appearance of clear-finished timber. If it is considered necessary to lock the cupboard, a three-leaf door is fitted instead of the roller blind.

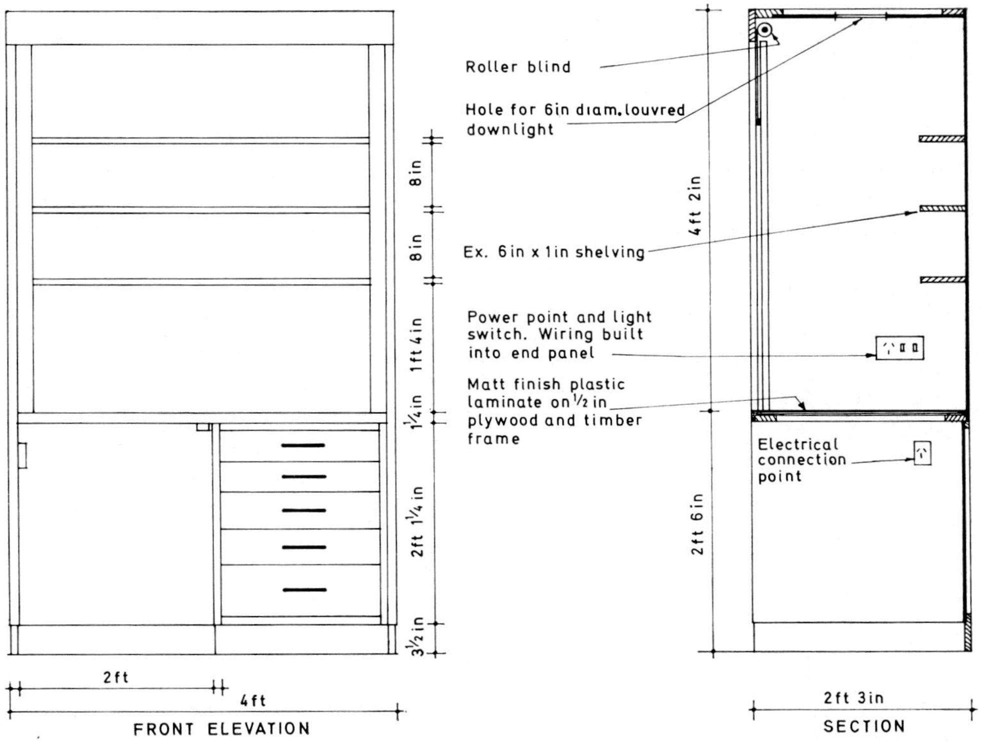

FIGURE 21 *CSIRO standard microscope cupboard.*

TITRATION BENCHES

Details of the CSIRO standard unit are shown in Fig. 22. As drawn, the top is at bench height, but occasionally table height is preferred.

SPECIAL BENCHES

Standard bench layouts can be adapted to meet most situations, but there are always exceptions which require individual treatment. The arrangement shown in Fig. 23 was designed to house five auto-analysers, and I am including a few details because the basic idea could very well be applied in other laboratories where there are a lot of instruments.

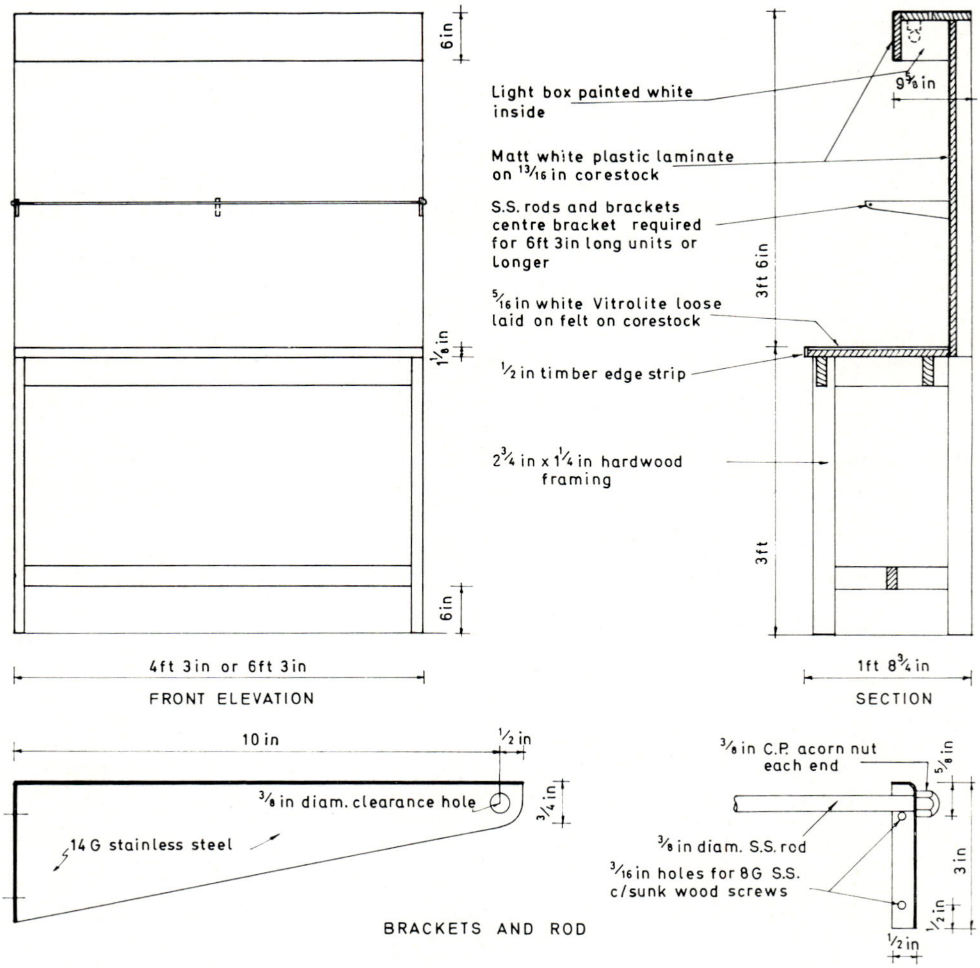

FIGURE 22 *CSIRO standard titration bench.*

FIGURE 23 *Special arrangement of benches, to provide easy access to backs of instruments, and keep the bench top free of electricity cables; CSIRO Cunningham Laboratory, Brisbane, Qld.*

In the laboratory illustrated, a space of 2 ft between two rows of benches acts as a service corridor in miniature; it provides easy access to the backs of the instruments and it keeps the bench top clear of power cables and tubing. The reagent bottles are grouped at one end of the unit, below bench level; they stand in a polythene tray and they are connected to the auto-analysers by tygon tubing with two-way teflon taps; as well as helping to keep the bench top clear, this also eliminates some toxicity problems.

Chapter 5

Services

An engineer once told me that a laboratory is just a building to house a lot of pipes—and he was quite right, up to a point! It is certainly true that pipes for services to benches, and ducts for air-conditioning, are a major portion of the work involved in planning a modern laboratory.

MECHANICAL SERVICES—DESIGN AND LOCATION

In some laboratories only hot and cold water, burner gas and compressed air might be reticulated; in others there might be hot and cold water, burner gas, compressed air and vacuum, and also restricted reticulation of demineralised water, cooling water and two or three bottled gases. Fume cupboards can be very few, or very many. It follows, therefore, that there is a tremendous variation in the amount of space required for pipes and ducts.

In the simplest type of installation, *i.e.* a two-storey laboratory with very few services, the pipes are accommodated in vertical ducts at 10 ft centres, and duct space for the fume cupboard exhausts is on the first floor. Additional floor-to-ceiling height on the ground floor will provide for the air-conditioning ducts; on the first floor, they would be in the roof space.

In its most elaborate form, there is a service floor between each laboratory floor; this means that, with the space required for vertical ducts, the area occupied by the mechanical services is actually greater than that devoted to research. The Salk Institute for Biological Studies, San Diego, USA was designed on this principle; the service floor is contained within the structural system which spans each of the laboratory floors, as shown in Fig. 24. It is interesting to note that the head room of 7 ft 6 in was considered by some to be extravagant; yet, in some other laboratories where the service floor principle has since been adopted, a height of 6 ft has been found insufficient. In an effort to reduce the height to an absolute minimum, designers have even considered the idea that the maintenance men should be supplied with wheeled trolleys so that they could work in a horizontal position!

Service mains should be accessible for their entire length from the point of entry into the building. They can be located below the ground floor slab (which, if necessary, should be raised to provide adequate working space), in a service tunnel below the ground floor, on the ceiling of the basement or ground floor, in the roof space or penthouse, or in a service corridor between laboratories. The design should incorporate the following:

The cold water service should be a ring main. The recommended pressure for a water pump is 50 lb/in^2; the water service should therefore be designed to maintain this pressure ± 10 lb. This requires an elevated tank, pump or reducing valve; a tank has the advantage that water

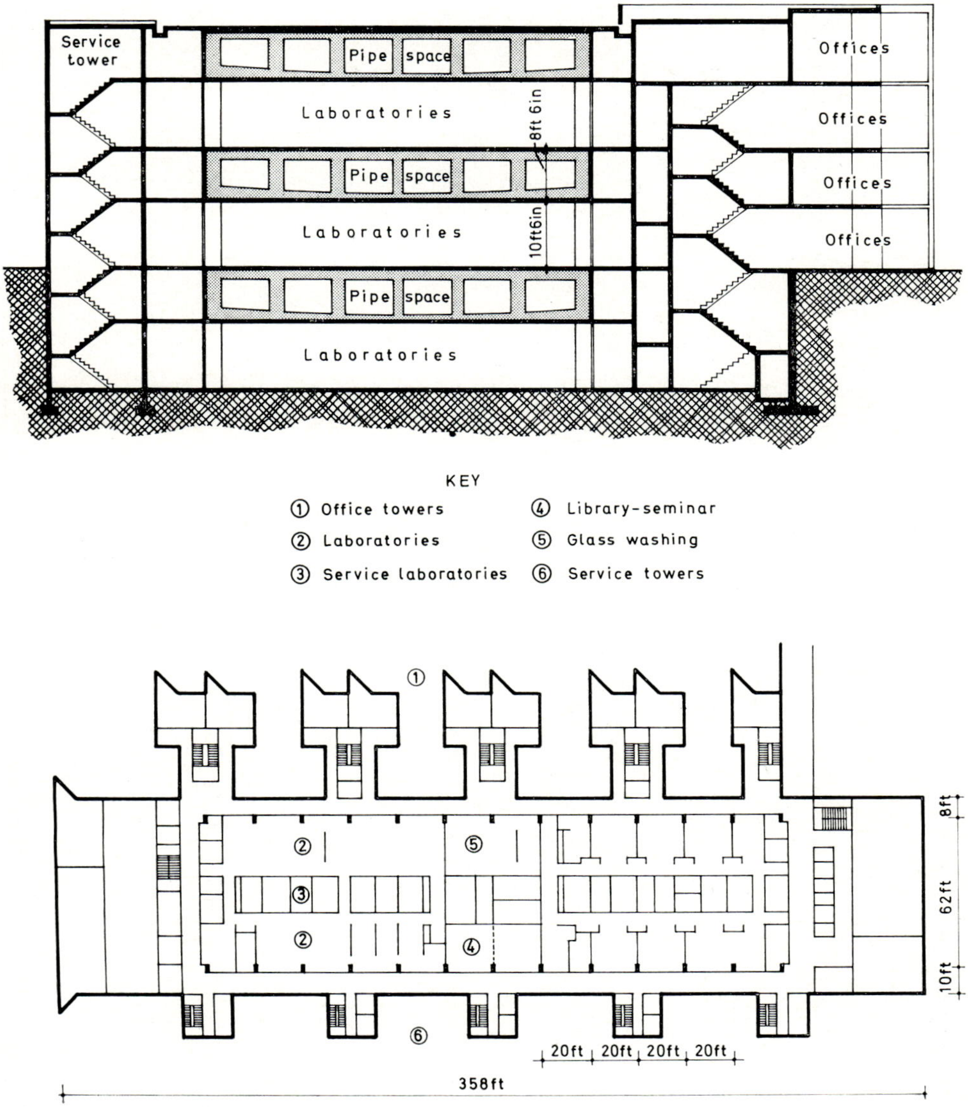

FIGURE 24 *Typical floor plan, and section showing service floors; Salk Institute for Biological Studies, California, USA.*

is always available for important experimental work, even if there is a breakdown in the supply.

In general, it is a good fault for the cold water pipes to be oversized, but care should be taken to ensure that all other pipes are not over-capacity. It is the project officer's responsibility to give the designing engineer details of the estimated consumption.

When selecting pipes and isolating valves, the deciding factors should be long life and minimum maintenance, rather than low initial cost.

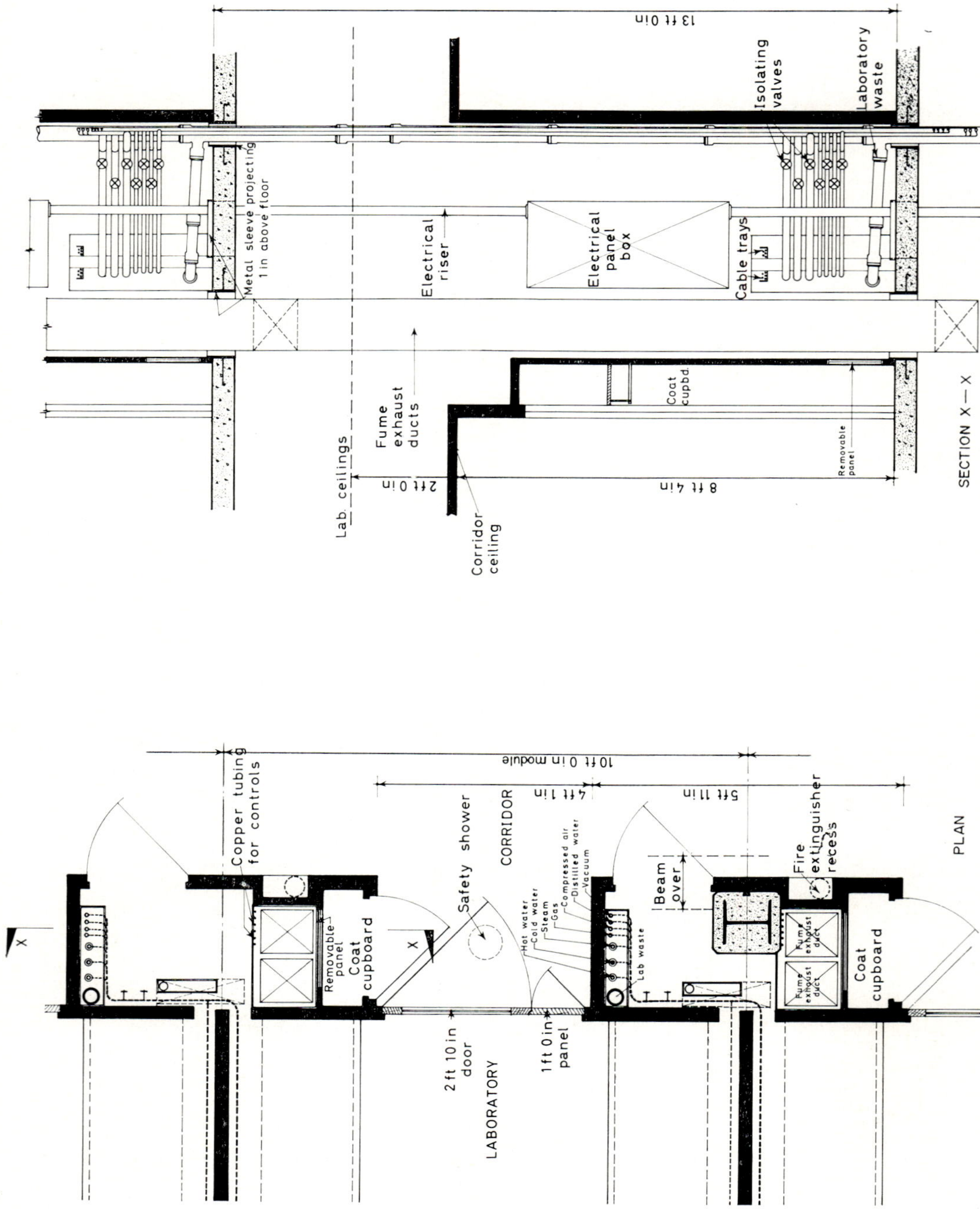

FIGURE 25 *Plan and section showing corridor service shafts; Homer Research Laboratories, Bethlehem Steel Corp., Pennsylvania, USA.*

(*a*) *Laboratory bench showing prefabricated panels, isolating valves, and construction of tiled bench top.*

(*b*) *Pipes and ducts reticulated in space above false ceiling.*

FIGURE 26 *Hoechst Research Centre, Frankfurt (M)-Hoechst, West Germany.*

All branch lines should be fitted with isolating valves to facilitate maintenance and alterations.

To provide for future extensions, there should be plugged tees, or tees with plugged isolating valves.

All piping should be colour-coded.

Many designs have been developed to reticulate pipes around buildings, but most are based on one of the following.

External wall

A vertical duct is provided at each module, as shown in Fig. 1. Above the bench, a removable cover extends to the ceiling; below the bench, a knee space provides access. One variation is to have vertical ducts in dummy columns at intervals of two or three modules, with horizontal pipe runs either exposed or in ducts with removable covers.

Corridor walls

A series of shafts for vertical pipes and ducts are formed by having two walls, 3 ft apart, between the laboratories and the corridor, as shown in Fig. 25. In some laboratories, a portion of this space is also used for fume cupboards (*see* Fig. 53).

Space above false ceilings

This provides excellent flexibility and is preferable to floor ducts (*see* Fig. 26); however, any trouble or maintenance will disrupt work in the laboratory below. The logical development of this system is the service floor (*see* p. 64 and Fig. 24).

Service corridor

This is a space separating two rows of laboratories and normally extending the full length and height of the building; it contains all the vertical and horizontal pipes and ducts. As mechanical services—particularly air-conditioning—continue to increase, a multiplicity of small ducts becomes less and less efficient; it seems, therefore, that the service corridor is not only the most practical but also the most economical solution for large projects.

In Australia, there is room for improvement in the design and installation of mechanical services, and the same could be said of electrical services. The chief reason is that each laboratory is designed on a 'one-off' basis; instead of developing a design and repeating it, a fresh start is made each time and consequently mistakes continue to be made. One of the most outstanding features of the Hoechst Laboratory at Frankfurt, and some of the pharmaceutical laboratories at Basle, is the meticulous attention to detail; this perfection has been achieved by repeating the identical laboratory layout in several buildings and progressively eliminating defects.

ELECTRIC LIGHT AND POWER

In many laboratories it seems that, as each year passes, more and more electricity is being used. The result is that switchboards are at near capacity within a few years of the buildings being occupied. The main switchboard should have generous spare capacity and provision should be made for both it and the transformer to be extended.

A well-designed installation for electrical power should incorporate sub-boards to serve not more than four laboratories (in some cases, individual laboratories) and each should have spare capacity; furthermore, the wiring should be in surface-mounted ducts. The position of power outlets for benches and fume cupboards is discussed on p. 92; double outlets should be used where possible and it is usual to provide one 15 A outlet for every fume cupboard. It is a wise safety precaution to use fittings which have a pilot light, even if they do cost slightly more. It seems that in some laboratories (*e.g.* electronics) there is almost no limit to the number of power outlets required; in such cases, it is a good idea to have a plug-board consisting of 10 outlets and a circuit breaker mounted on a panel which can be plugged into one wall outlet.

In Australia, laboratory lighting is generally designed to give 30 and occasionally 50 lumens per square foot. For an interesting comment about lighting intensities in the United States of America, *see* p. 135.

NUMBER AND POSITION OF OUTLETS

The number of service outlets to be provided on a bench requires close scrutiny; the aim is to be generous, but not extravagant. In some types of laboratories, adequate services on a 15 ft peninsular bench would be: 4 cold water over a central trough; 1 gas; 2 air; 4–6 power outlets on each side; 1 hot and cold water, and 2 cold water, over sink at end of bench.

In assessing the number of outlets required, it is helpful to consider what has been done in existing laboratories; for this reason, details of service outlets have been included in the descriptions of Australian and overseas laboratories given in Chapters 7 and 8. The Nuffield Foundation conducted a comprehensive survey of the use of hot and cold water, gas and power by scientists and assistants in various types of laboratories and this valuable information is set out in Chapter 3 of their publication *The Design of Research Laboratories*.

It is not easy to obtain factual information on this subject and, in order to find out just how much the service outlets provided were actually used, I arranged for a restricted survey to be taken in eight CSIRO laboratories. The recordings were made by an independent observer during six 1-hour periods (at varying times of the day) at intervals of 4–6 weeks and the results given in Table 2 show the total usage during the survey time of six hours. These data indicate that the number of cold water and vacuum outlets could be reduced; there are far too many gas, but there are about the correct number of hot water, air and power. Nowadays gas is being used less and, in many laboratories, it could be omitted entirely or at least be reticulated on a very restricted basis.

Service outlets are mounted on the service strip; this is at the back of a wall bench, or in the centre of a peninsular bench. An alternative position is on the reagent shelf unit, either on the rail under the lower shelf or on the panels which support the shelves at about 5 ft centres. The use of the reagent shelf for this purpose keeps the bench space clear and also reduces on-site work because the unit, complete with pipes and flex, can be assembled in the factory. A standard grouping of outlets simplifies installation and therefore reduces cost.

WASTE PIPES

When selecting the type of waste pipes to be used in a job, the factors to be considered are resistance to acids, alkalis and solvents, the effect of hot effluent over long periods and the cost of installation and maintenance. In general, one tends to exaggerate the severity of the conditions; remember that, in the past, even galvanised wrought iron pipes have given years of satisfactory service in some laboratories.

Glass is used extensively in the United States of America at present, but plastics are generally used in most other countries. In Australia in 1971, rigid PVC pipe and polypropylene bottle

Table 2 Use of laboratory services during a restricted survey of CSIRO laboratories

Test lab	Area ft²	Occupants	Cold water: No. outlets provided	Cold water: No. used	Cold water: Frequency of use	Hot water: No. outlets provided	Hot water: No. used	Hot water: Frequency of use	Gas: No. outlets provided	Gas: No. used	Gas: Frequency of use	Compressed air: No. outlets provided	Compressed air: No. used
A	400 + office	1 RO 1 EO 1 Asst.	33	7	2 constantly 3 to vac. pump 2 occasionally	3	2	constantly	12	4	constantly	2	1
B	330 + office	1 RO 1 TO	33	6	3 constantly 1 to vac. pump 2 occasionally	2	2	occasionally	12	1	occasionally	1	—
C	330+ office	1 RO	23	5	2 constantly 1 to vac. pump 2 occasionally	2	2	constantly	8	1	occasionally	2	—
D	280 + office	1 RO 1 Asst.	26	5	2 constantly 1 to vac. pump 2 occasionally	2	2	constantly	12	1	on 1 occasion	2	1
E	500 + office	2 EO 1 Asst.	33	8	2 constantly 3 to drip feed 1 to still 2 occasionally	2	2	constantly	16	—		9	3
F	430	1 RO 1 EO 1 Asst.	26	6	2 constantly 2 to drip feed 1 to still 1 occasionally	2	2	1 constantly 1 occasionally	14	1	on 2 occasions	8	2
G	750 + 2 offices	2 RO 2 EO 2 Assts.	41	10	3 constantly 4 to drip feed 3 occasionally	3	2	constantly	38	1	on 1 occasion	11	2
H	490 + office	2 EO 1 Asst.	24	6	3 constantly 1 to drip feed 2 occasionally	2	2	constantly	20	4	occasionally	3	1

traps are common; they offer good resistance to acids and, providing they are flushed down with water, they are not affected by organic solvents or other organic compounds in the quantities normally used in laboratories. PVC is thermoplastic and therefore hot effluent causes softening and expansion; it also moves significantly with variations in ambient temperature. Polypropylene traps are normally fitted with polypropylene bowls, but glass bowls are available at slightly higher cost. The latter have better resistance to solvents and, being transparent, they simplify maintenance, especially where mercury is being used. Increased resistance to solvents can be obtained by using polyethylene pipe and, if this is considered to be inadequate, then glass should be used.

Glass waste pipe and fittings are available in Australia, but they are at least double the price; for this reason, they have not been used extensively as yet. The glass is tempered and it is completely resistant to all chemicals—it will even withstand hydrofluoric acid for long periods; it has a low expansion rate and is unaffected by hot effluent; it has good shock resistance and maintenance is almost nil. Glass is assembled with a compression joint consisting of a teflon liner, gasket and stainless steel clamp.

Some of the more important points in the design and installation of PVC pipework are:

$1\frac{1}{2}$ in diameter outlets to drip cups, and 2 in diameter to troughs and sinks.

Most authorities require every trough and sink outlet to be fitted with a trap, but they will permit up to three drip cups to be connected to one trap.

Sinks on either side of a partition should have separate waste pipes.

Table 2—*continued*

	Vacuum			*Steam*			*Power*			*Fume cupboards*			
Frequency of use	*No. outlets provided*	*No. used*	*Frequency of use*	*No. outlets provided*	*No. used*	*Frequency of use*	*No. outlets provided*	*No. used*	*Frequency of use*	*No. provided*	*Frequency of use of fan*	*OFFICES Average use—minutes per hour*	*WRITING SPACES Average use—minutes per hour*
continuously on 1 occasion	—			—			24	24	14 continuously	1	continuously	15	8 each
	—			—			26	20	10 continuously	1	continuously	16	11
	—			—			18	14	8 constantly	1	continuously	10	Not used-only one occupant
on 1 occasion	—			—			20	18	10 constantly	1	continuously	35	Nil
on 1 occasion	29	6	constantly	2	—		34	28	15 constantly	2	continuously	Section leader	12 each
on 1 occasion	18	5	constantly	1	—		29	23	13 constantly	2	continuously		13 each
on 1 occasion	23	9	5 constantly	2	—		43	38	18 constantly	3	continuously	35 12	Nil
on 3 occasions	8	3	constantly	2	—		32	26	12 constantly	3	continuously	Section leader	18 each

All horizontal pipes must be set to a fall. The normal grades for self-cleaning are 1 in. in 20 in for a 2 in diameter pipe, and 1 in. in 25 in for a $2\frac{1}{2}$ in diameter pipe. Laboratory waste does not carry solids and therefore most authorities will agree to reducing this fall, for example, 2 in diameter pipe from 1 in. in 20 in to 1 in. in 60 in. Regulations state that the maximum length of a graded unvented pipe is 20 ft; but again, depending on the number of fittings, the authorities will usually allow this to be increased to about 26 ft.

Back venting can generally be eliminated entirely by the simpler and less costly expedient of using oversized pipes; thus a 2 in diameter waste from a sink at the end of a peninsular bench is increased to $2\frac{1}{2}$ in from the point at which the second and subsequent sinks are connected. Four $2\frac{1}{2}$ in diameter graded wastes, *i.e.* two on each of two floors, can be connected to a 3 in diameter stack. Above bench level on the top floor the stack is reduced to 2 in diameter and carried up to a height of 2 ft above the roof. Details of a typical layout are shown in Fig. 27.

The underground pipe from the stack to the neutraliser is 4 in or 6 in diameter earthenware, with spigot and socket joints sealed with a $\frac{9}{16}$ in diameter teflon ring.

When assembling pipes and fittings, particular care should be taken to establish proper grade and alignment before jointing; they must not be stressed by straining them into position.

Any pipe passing through a floor or wall should have a clear space of at least $\frac{1}{2}$ in around it.

Pipes should be at least 3 in clear of heating pipes and conduits.

Solvent welding (as approved by the authorities) is recommended for jointing.

Expansion joints should be fitted to the stack at each floor at the downstream end, and to

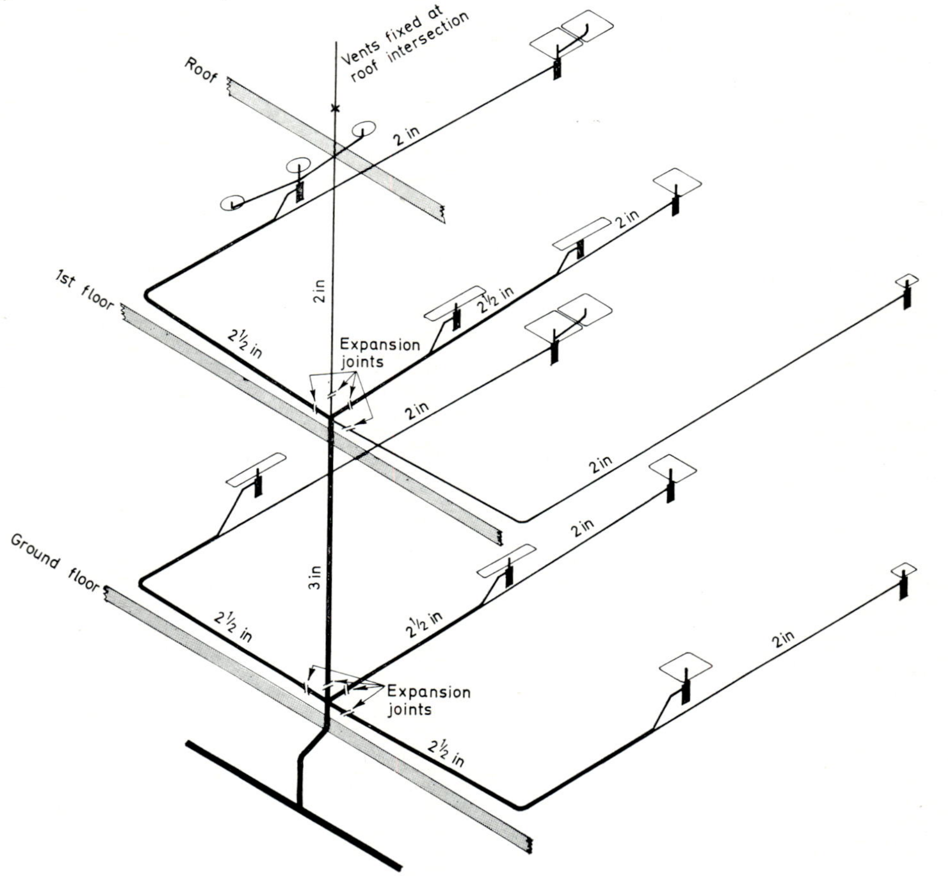

FIGURE 27 *Diagram showing a typical layout of waste pipes in a two-storey building.*

graded pipes more than 4 ft and less than 20 ft in length. Care should be taken that the pipe is not bottomed in the expansion joint. The space to be allowed for expansion should be:

Cold weather (up to 50°F) — $\frac{3}{4}$ in   Warm weather (51 to 80°F) — $\frac{5}{8}$ in
Hot weather (above 80°F) — $\frac{1}{2}$ in

Graded pipes should be supported at intervals not greater than 4 ft, with clips so placed as to maintain the pipes on a uniform grade; vertical pipes to be supported at intervals not greater than 6 ft. Pipe clips, when completely tightened, should permit longitudinal movement.

NEUTRALISERS

The effluent from laboratory sinks contains varying quantities of acids, alkalis, and solvents and sewerage authorities require that it be neutralised before entering the mains. If there are only a few sinks and they are some distance apart, individual neutralisers are the simplest and most economical; they can be vitreous china, PVC or stainless steel. However, it is usual to connect all the sinks to one or more large neutralisers.

Some authorities approve a retention or buffer tank which dilutes and self-neutralises, but most require that the effluent be neutralised by soda-ash; in some situations, marble chips can be used.

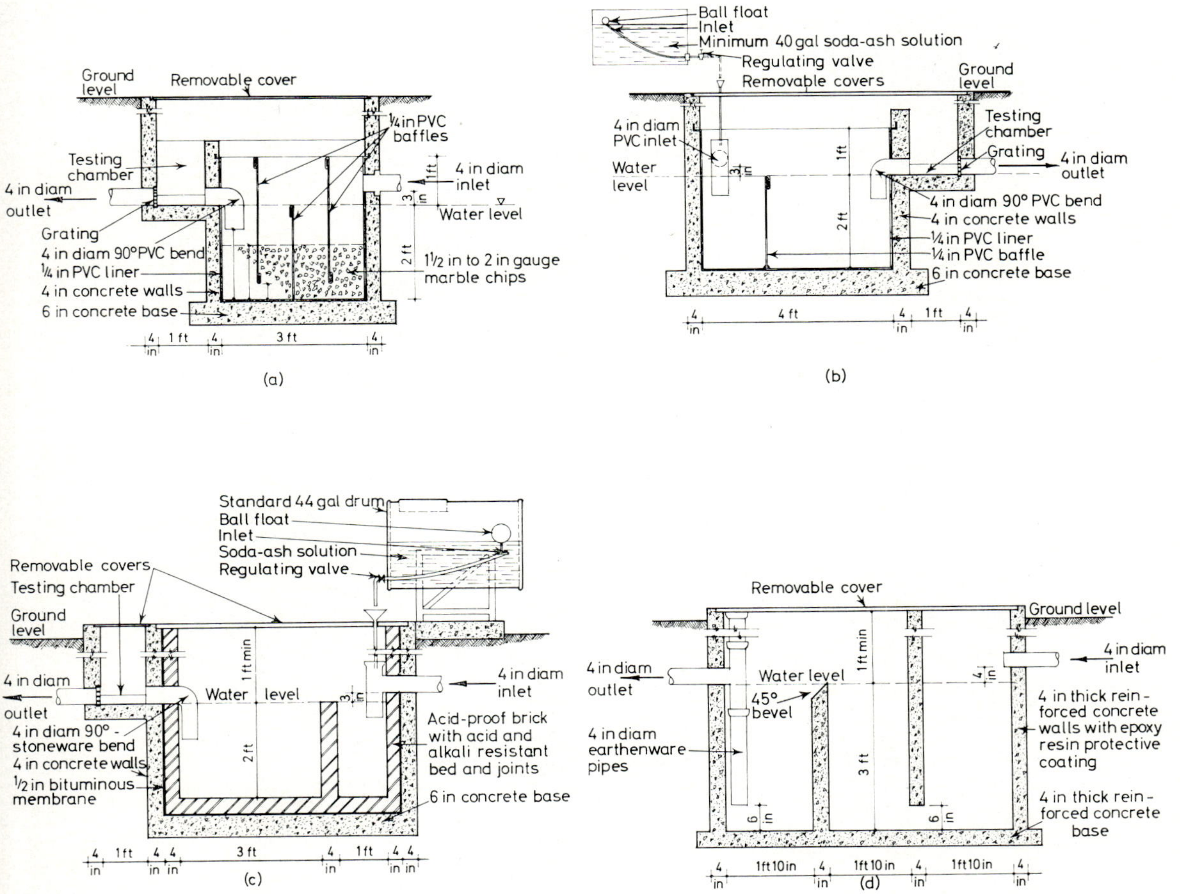

FIGURE 28 *Four types of neutralisers:—(a) PVC lining and marble chips; (b) PVC lining and soda ash; (c) acid-proof brick lining and soda ash; (d) self-neutralising tank of concrete with protective coating of epoxy resin.*

Details of these three types of neutralisers are shown in Fig. 28. If marble chips are used, the size of the chips is important because the neutraliser will quickly become clogged if they are too small; 1½–2 in is recommended. If large quantities of sulphuric acid are being discharged, marble chips are ineffective because they become coated with calcium sulphate and the chemical action ceases. With soda-ash neutralisers it is usual to have a manually controlled drip feed, but an automatic pH controller with mechanical stirrer and feed can be used in large installations (*see* p. 103).

FIGURE 29 *PVC linings for neutralisers, showing reinforcing fins; on the job, these will be encased in concrete. (By courtesy of P. C. Ridgeway & Co. Pty Ltd, Carlton, Vic.)*

The construction of a neutraliser which will meet the approval of the authorities varies; some will accept a reinforced concrete tank painted with epoxy resin, others require PVC lining. The PVC tanks are reinforced by 3 by $\frac{3}{4}$ in fins at about 1 ft 6 in centres, as shown in Fig. 29. Whilst PVC is satisfactory in most cases it can give trouble with expansion if the effluent is hot, and it will soften and bulge if there is a build-up of solvents in the neutraliser; it will be attacked by light solvents at the water line and by heavy solvents at the bottom. For large installations and severe conditions, the concrete tank should be lined with acid-proof bricks set in acid-resistant mortar, as shown in Fig. 28.

The capacity of the neutraliser is determined not by the number and total capacity of the sinks connected to it, but by the estimated maximum number of gallons per hour which will be discharged into it. If marble chips are to be used, and assuming a discharge of 200 gal/h, a retention time of 45 min and chips occupying three-fifths of the volume, then the capacity of the neutraliser (to the top of the baffle) is

$$200 \times \frac{3}{4} \times \frac{5}{2} = 375 \text{ gal}$$

If soda-ash is used, the retention time is only 15 min and the total capacity is therefore 50 gal.

It is not surprising that a great deal of slime, sediment, and even algae, collect in neutralisers. It is essential, therefore, that they be regularly and thoroughly cleaned.

FUME CUPBOARD EXHAUST SYSTEMS

No matter how well a fume cupboard is designed, it will not function efficiently unless its exhaust system is also well designed. Fumes should be ducted to the roof by the shortest possible route, and they should be discharged in such a way that they disperse well away from the building concerned and any others in the immediate vicinity.

All this probably sounds too obvious, but it is a fact that very often the fume cupboard exhaust system is hardly thought about until after the design brief and preliminary plans and elevations have been prepared. At this stage it is too late to do anything other than try to fit it in the best way possible and this is just not good enough! With an item as important as the disposal of corrosive and dangerous fumes, one should not be forced to compromise and perhaps have to accept such highly undesirable features as:

- long horizontal runs of ducting;
- too many bends and junctions which, to make matters worse, are quite often badly designed;
- grouped outlets with individual fans, either exposed or screened by parapets;
- outlets which discharge the fumes behind louvres in a roof lantern;
- poor accessibility to fans for maintenance.

In Australia, at least, there are defects in the design and construction of many fume cupboard exhaust systems. In laboratories where the cupboards are used only occasionally, and then only for light work, these faults are relatively unimportant, but when the cupboards are in constant use or the conditions are severe, even seemingly small defects can become serious. The pity of the situation is that most of these problems need not arise if the design is given the consideration it warrants right at the start.

Exhaust fans

Exhaust fans should be located near the point of discharge rather than immediately above fume cupboards; they should therefore be in the roof space or penthouse, or on the flat roof. The duct between the cupboard and the fan is then under negative pressure and, if there are any holes in the ductwork, air is drawn into the duct instead of fumes being blown out into the building; furthermore, the noise level in the laboratory is reduced to a minimum.

FIGURE 30 *Flue-mounted fume cupboard exhaust fan; it has a PVC impeller.* (*By courtesy of Conditionaire Engineering Pty Ltd.*)

In buildings which are not air-conditioned it is usual to have one exhaust fan to each fume cupboard; individual control is generally much more satisfactory—the fan can be switched on as required and dampers are not necessary. However, if it is operating continuously, the ducts from several fume cupboards (preferably in the one laboratory) can be connected to one fan.

A range of centrifugal fans specially designed to exhaust corrosive fumes is available commercially in Australia. The impeller is non-welded and can be made from PVC, aluminium or stainless steel. PVC is used for general fume cupboard work, especially where acid fumes are to be discharged; aluminium and stainless steel are used for special work, for example radioisotopes and some solvents. The PVC and aluminium blades are extruded and inclined backwards; a forward-inclined impeller is available for cases where the resistance of the system is high. The shaft is PVC-coated steel and both direct and belt drives are used; the belt has the advantage that, within certain limits, the speed and capacity of the fan can be varied by changing the pulley diameter. The fan housing is compressed asbestos cement and No. 316 stainless steel. The simplest and most economical installation is the flue-mounted fan shown in Fig. 30; however, the pedestal-mounted type with vertical discharge is recommended for general use (*see* Fig. 31).

The use of two-speed motors should always be considered. Whilst the fume cupboard will be designed so that it operates satisfactorily and at an acceptable noise level when the motor is at low speed, it is a decided advantage to be able to switch to high speed for the rapid exhaust of

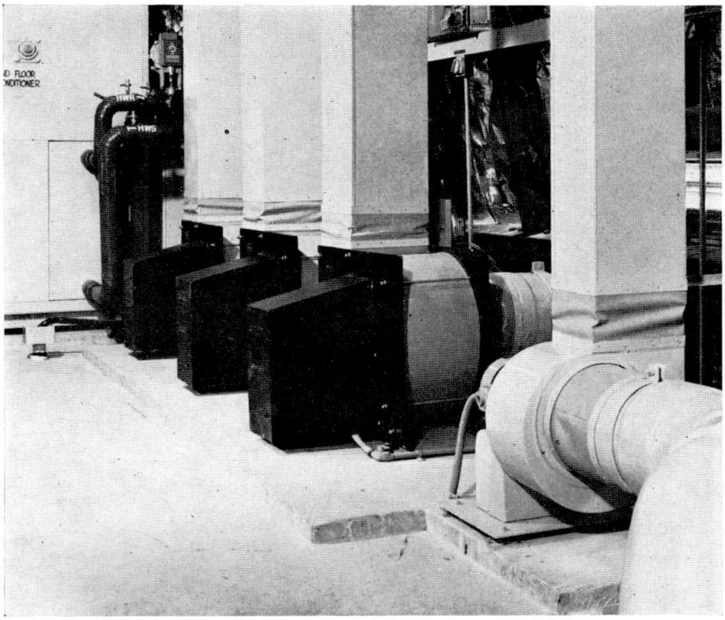

FIGURE 31 *Pedestal-mounted fume cupboard exhaust fans with vertical discharge—installed in the Australian Mineral Development Laboratories, Frewville, SA.*

dense or dangerous fumes. At 0·25 in w.g., a 12-in diameter fan coupled directly to a $\frac{1}{2}$-hp continuously-rated two-speed motor will exhaust 700 ft^3/min at 960 rev/min and 1300 ft^3/min at 1440 rev/min. If two-speed fans are used in air-conditioned buildings, the system must, of course, be designed so that the air balance is maintained. Two-speed motors have been used to reduce the quantity (and therefore the velocity) of air being exhausted when the sash is down; a trip switch is operated by the movement of the sash.

If there is a vertical discharge, a drainage outlet should be provided to remove rainwater which may enter and lodge at the bottom of the fan housing when the fan is not operating. These outlets are connected to the waste system by $\frac{3}{4}$-in diameter polythene or PVC pipe. However, when the fan is operating, most of the rain will be blown aside because the maximum velocity of large raindrops is 1800 ft/min which is about the same as the velocity in the stack.

With a vertical discharge it is sometimes necessary also to have an anti-downdraft device. This consists of lightweight flaps fitted to the top of the stack; they are opened by the velocity of the exhaust air when the fan is operating and they drop back into position when the fan is turned off. I quote one example where this device should have been fitted: In a building which was centrally heated, the exhaust fans were flue-mounted 2 ft 6 in apart in several groups on the flat roof. Apparently there was a negative air pressure in a laboratory two floors below; the result was that, when the fume cupboard in this laboratory was not operating, fumes being discharged from a nearby stack were drawn along about 50 ft of duct into the room.

Ducts

Properly designed ductwork is an essential part of an effective fume exhaust system. To obtain minimum resistance and a streamlined airflow, horizontal runs should be avoided or reduced to an absolute minimum and right-angle bends and junctions should not be used.

If the velocity in the duct is too high, it will cause objectionable noise; nevertheless, it is desirable for it to be on the high side because it gives better dispersal of fumes and there is less risk of condensation. As a guide ducts are generally 9 in diameter for cupboards less than 6 ft long and 12 in diameter for cupboards 6–8 ft long.

The duct extends from the top centre of the fume cupboard to the exhaust fan; if it is not central there will be an unequal distribution of air over the face opening. Of course, there is always the exception. I have seen a 9-ft fume cupboard with the exhaust duct at one end; an equal air distribution was obtained by adjusting the space between the four sliding baffles at the top.

Materials used for ductwork are asbestos cement (with two coats of chlorinated rubber inside), PVC, galvanised iron (with two coats of tar epoxy inside) or stainless steel; all have good resistance to corrosive fumes. The fire authorities prefer metal. Asbestos cement has the advantage that, because of its extra weight, it will dampen noise, but it will shatter in a fire. PVC has the advantage that it can be fabricated to any shape or size, but the disadvantage that it will burn, although it is self-extinguishing. Heavy-duty flexible PVC ducting is very useful for connections in awkward or restricted positions.

Flexible connections

A flexible connection isolates vibration and reduces the noise transmitted through the ductwork to the laboratory. One should be fitted on the inlet side of the fan and, if considered necessary, there should be another on the outlet side. A suitable material for these connections is flexible PVC or polypropylene.

Discharge of corrosive fumes

When a cowl is fitted over the stack fumes are discharged below the horizontal and, on a windy day, they are drawn into the turbulent air around the building; even on a still day they tend to drift downwards. It will be seen, therefore, that fumes can easily be drawn back into the building through air inlets or windows and I could quote instances of this happening—even when the points of discharge and re-entry have been as far apart as 100 ft horizontally or three storeys vertically. The following points are clear:

to obtain maximum dispersion of corrosive and dangerous fumes, they should be discharged vertically;

the position and height of outlets should be considered in relation to the prevailing wind and the likely pattern of the airflow around the building;

the positions of air inlets require individual consideration, but, in general, they should not be on the leeward side of the building.

Some novel attempts have been made to improve the dispersal of fumes. The funnel of the liner Queen Elizabeth II has been designed so that the base of the plume remains parallel to the waterline; an upward current of air is created by the use of side baffles and a scoop-shaped base to the stack. In his book *Air Pollution*, R. S. Scorer points out that an aerofoil fitted along the top edge of a building (*see* Fig. 32) has the effect of producing a shallow jet of air across the rooftop; thus turbulence is eliminated and the dispersal of fumes is greatly improved.

Treatment of fumes

The quantities of corrosive and toxic fumes discharged from laboratories are very small compared with those from some industries. Nevertheless, the current alarm about air pollution may lead to the introduction of regulations requiring the treatment of the more severe fumes from some laboratories. With the materials now available it should be a relatively simple matter to develop a modern version of a scrubber tower; already a small unit consisting of a polypropylene filter-pad and water spray is on the market.

Airflow around buildings

In recent years one has often read about unusual behaviour of rain and wind in the vicinity of tall buildings. In moderate or high winds fumes discharged vertically from short stacks become completely dominated by the turbulence caused by the building, and they may be dispersed so violently as to envelop the whole building except for the windward side.

Information on this subject is limited, but some investigations have been made both in the field and in wind tunnels; details are given in two reports—'Studies of Air Flow Round Buildings' by Wise, Sexton and Lillywhite of the Building Research Establishment, Great Britain, and 'Gas Diffusion Near Buildings' by James Halitsky, New York University. The former gives observations and results of wind tunnel tests that are typical of conditions around parallel rows of similar models (*see* Fig. 32), and it states*:

> 'The long arrows represent streamlines and show the way that buildings can cause change of direction of the airstream. A parallel airstream passing over open country is deflected some distance before reaching the first building and a vortex forms at the windward face of the building. The flow lifts over the windward edge of the roof and then descends, a vortex occurring in the space between the buildings; this pattern is repeated with succeeding buildings and a turbulent wake forms on the leeward side of the downstream building. . . . The stream of air which is deflected upwards by the leading edge of the roof on the windward building clears the central building and descends on the downstream building.'

Height of stack

The height from the roof to the top of the stack varies from a few inches in some laboratories to a few feet in others. The question arises—what height should the stack be to achieve satisfactory dispersion of fumes?

In his report 'Estimation of Stack Height Required to Limit Contamination of Building Air Intakes', Dr Halitsky outlines a procedure to determine the minimum stack height consistent

* Reproduced by permission of Her Britannic Majesty's Stationery Office, London.

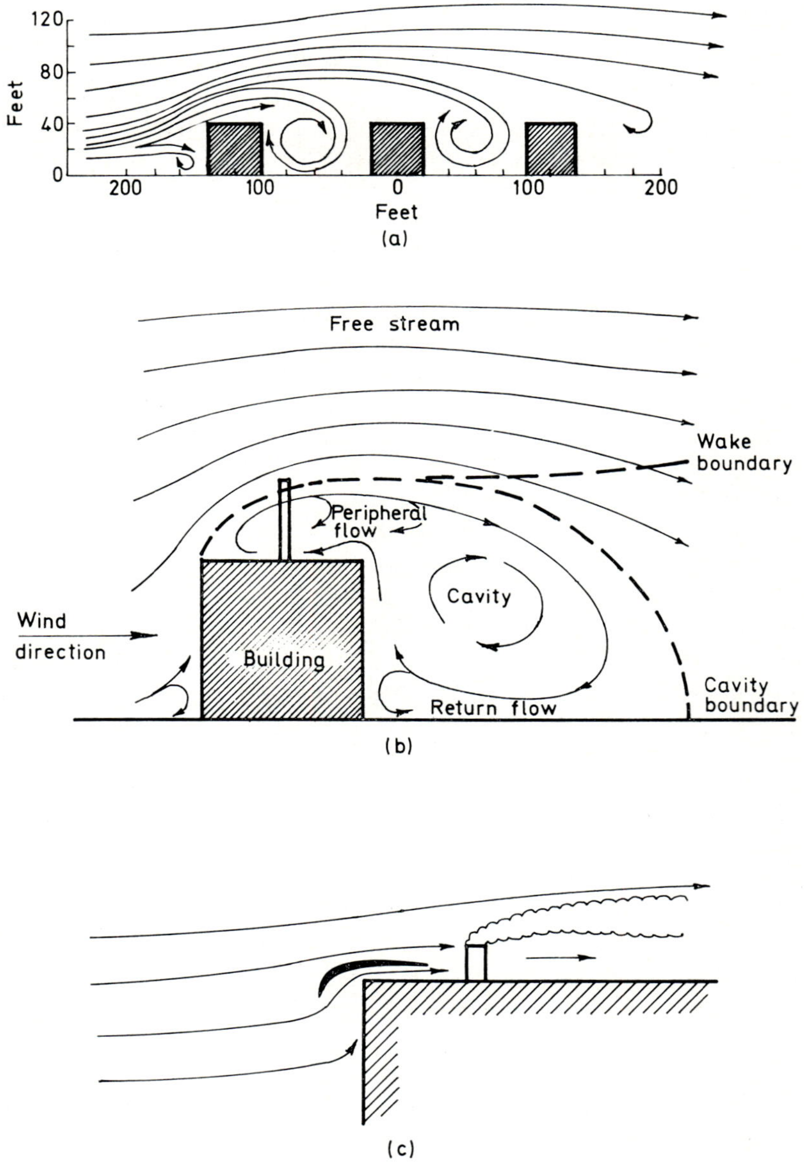

FIGURE 32 *Diagram showing (a) direction of air flow across three parallel buildings, (b) height of stack required to discharge fumes above the zone of turbulent air, and (c) aerofoil to reduce turbulence, fitted along the top edge of a building.*

with a given maximum allowable gas concentration in the cavity, and he calculates an example:

> 'The building is a 50 ft high cube with fresh air intakes in the side walls. A continuous process within the building generates 10 ft^3/min of pure gas which must be discharged through the centre of the roof. The design wind speed is 10 mph. The maximum allowable concentration of the intake is one part per million. How high should the point of discharge be above the roof if the gas is released in a fume hood handling 2400 ft^3/min and discharging at a velocity of 2400 ft/min?'

Dr Halitsky concludes that it should be 18·4 ft.

It is obvious that the height of outlets above the roof will have to be considerably increased in order to get the best results. In their efforts to achieve a neat appearance, many architects have concealed fume cupboard exhausts; indeed, if fume dispersion and the possible contamination of air inlets has been considered at all, it has been regarded as being of less importance than aesthetics. In laboratories where there are a lot of fume cupboards, a solution is to connect the ducts to one or more tall stacks so that, instead of concealing the outlets, they become a prominent and integral part of the design.

AIR-CONDITIONING

Despite its high cost, air-conditioning is regarded as essential in every modern laboratory; it gives positive ventilation and filtered air at constant temperature. The design of air-conditioning systems for laboratories is complicated by recirculation, fume cupboards and heat load, and these are discussed below.

Recirculation

In office buildings it is usual to recirculate about 75% of the air and this can also be done in some laboratories, for example, physics. However, in many laboratories (chemistry, etc.) this is not acceptable because of the risk of distributing toxic or solvent fumes throughout the building. Obviously a 'once-through' system is expensive and therefore it should be restricted to the areas where it is absolutely necessary. Depending on the size of the project, administrative space can be provided in a separate building, or be grouped on one floor or on one side of a floor; these areas will then have separate air-handling plants with provision for recirculation. Some air-conditioning systems have been designed on the basis of 'once-through' to laboratories with fume cupboards, but with recirculation to laboratories without fume cupboards.

If, instead of a central plant, each laboratory (or small group of laboratories) has its own air-handling unit, then recirculation is acceptable—even in some chemistry laboratories. A supply of fresh air (filtered and partially heated) is ducted to each laboratory, and CSIRO experience indicates that three changes per hour is sufficient to dilute odours to an acceptable level.

Fume cupboards

Fume cupboards present a major problem. In a small laboratory there might be 20 cupboards and in a large project more than 100, and each one can exhaust about 900 ft^3/min. The quantity of conditioned air exhausted by the fume cupboards can be reduced by providing an auxiliary air supply to each cupboard; this system is not without its problems and it is discussed on p. 50. It is probably best to give two examples which illustrate the effect of fume exhaust on the design of air-conditioning systems:

In the Monsanto laboratory in Missouri, USA, there are four fume cupboards in each of 80 laboratories and an analysis of the situation is given on p. 53. This is a 'once-through' system; conditioned air from the laboratory is exhausted through the four fume cupboards which operate continuously. The cupboards are designed to operate satisfactorily at 550 ft^3/min, and it was considered that 2200 ft^3/min was about the maximum quantity of air which could enter the laboratory without causing discomfort to the occupants.

A preliminary report on the design of the proposed CSIRO Division of Applied Chemistry Laboratories indicates that there will be 113 fume cupboards 8 ft long and 32 cupboards 4 ft long. The estimated cost of the refrigeration, heating and ductwork for 'once-through' air conditioning—based on a temperature difference of 45°F and a cost of $1400 per ton—would be:

volume of air exhausted: $(113 \times 900) + (32 \times 700) = 124\,100\ \text{ft}^3/\text{min}$

maximum cooling required: $\dfrac{124\,100 \times 1{\cdot}1 \times 25}{12\,000} = 284$ tons

cost: $284 \times \$1400 = \$397\,600$

Note: this does not include the cost of the 145 fume cupboards, the exhaust fans or services to the cupboards

The design of the air-conditioning system is simplified if the scientist will agree to have only his 'research' fume cupboards in the laboratory and his 'routine analysis' cupboards in another room which can be supplied with filtered (and, if necessary, partially heated) air, independent of the main air-conditioning system. As it is likely that only one-third of the 'research' cupboards will be used simultaneously, these can be fitted with individual two-speed exhaust units and the additional air required can be obtained by using the corridor as an air plenum.

Heat load

The heat gain due to the sun, temperature of the outside air and lighting can be readily calculated. It can be reduced if, in the preliminary stages of design, due consideration is given to such items as orientation, area of glass, external shading of windows and insulation. For laboratories, an occupant loading of one person per 120–150 ft^2 is adequate. To my knowledge, the amount of heat generated by equipment in various types of laboratories has never been investigated in Australia. The *ASHRAE Guide and Data Book, Applications*, 1968, Chapter 10, gives the information shown in Table 3; this means that, in a lightly loaded three-module laboratory having an area of 720 ft^2, the heat load at 15 Btuh/ft^2 is 3kW, or 1 kW per 24 by 10 ft module.

In some laboratories, the heat generated is so small that no provision is made for it in the design calculations, but in others (mostly electronics and instrument laboratories) it can be quite significant. Of the four laboratories with which I have been involved most recently, it was decided not to make any provision in the air-conditioning for equipment heat load in two of them; the other two were designed to allow for 1 kW per 150 ft^2. The scientist can do much to eliminate or reduce

TABLE 3. Details of heat gain from laboratory equipment.[a] (By courtesy of ASHRAE, New York, USA)

Equipment loading	*Typical occupancy examples*	*[b]Net internal heat from laboratory equipment, total Btuh/ft²*		*Sensible heat ratio*
		Base maximum[c]	*Peak maximum[c]*	
General use	Chemistry, Biology, Physics	15	30	0·8 –0·95
High	Physics, Physical chemistry	30	60	0·8 –0·95
Very high	Electronics	70	140	0·95–1·0

[a] Based on heat release for bench type laboratory equipment in a 250 ft² module. Large individual items should be evaluated separately.

[b] Includes net gain to space after deductions for diversity and hooding.

[c] Base maximum rates are design values for majority of spaces. Peak maximum rates are design values for few laboratory spaces with especially high concentrations of equipment and are usually handled by supplementary conditioning.

the amount of heat gain in the laboratory by placing equipment in a fume cupboard, under an exhaust hood or by using insulation or cooling water.

In the August 1964 issue of *Air Conditioning, Heating and Ventilating*, published by Industrial Press, New York, H. E. Twietmeyer, General Motors Research Laboratories, states:

> 'In our laboratory, based on typical equipment used in these activities and discussions with the research personnel about their present and future equipment needs, working habits and operational factors, physics and basic electronic laboratory design equipment loads were set at 15 Btuh/ft^2, and all other chemistry and special purpose laboratories at 50 Btuh/ft^2. A recent comparison of the original equipment list with the equipment currently being used in these laboratories shows differences, but the actual measured load at this time is reasonably close to the original design load. In other words, the changes through the years stayed within the design allowance.
>
> 'The connected load is the sum of all the name plate ratings of equipment located within a lab module. The name plate ratings were adjusted to real life heat loads by applying load, use and diversity factors.
>
> 'Examples of the equipment design loads and actual measured equipment loads are illustrated in Fig. 1 (reproduced in this book as Fig. 33). In spite of some deviations indicated in this recent study, we feel that the design load of 50 Btuh/ft^2 has been a reasonable selection for our laboratory operations during the past 10 years.'

Laboratories frequently need cooling water to carry away the heat generated by scientific equipment. If only small quantities are involved, the authorities quite often do not object to it being discharged into the waste system. However, as the trend is towards more equipment requiring cooling water, the matter becomes much more serious; for example, in 1960 the CSIRO

National Standards Laboratory was using 38 000 gal of cooling water each week and it was run to waste. The authorities insisted that a recirculating system be installed; it cost $15 000.

The installation of a recirculating system is necessary both for economy and conservation of water; it also has the very definite advantage of providing a constant supply of clean water at a controlled temperature and pressure, and it reduces the air-conditioning load. It is essential to

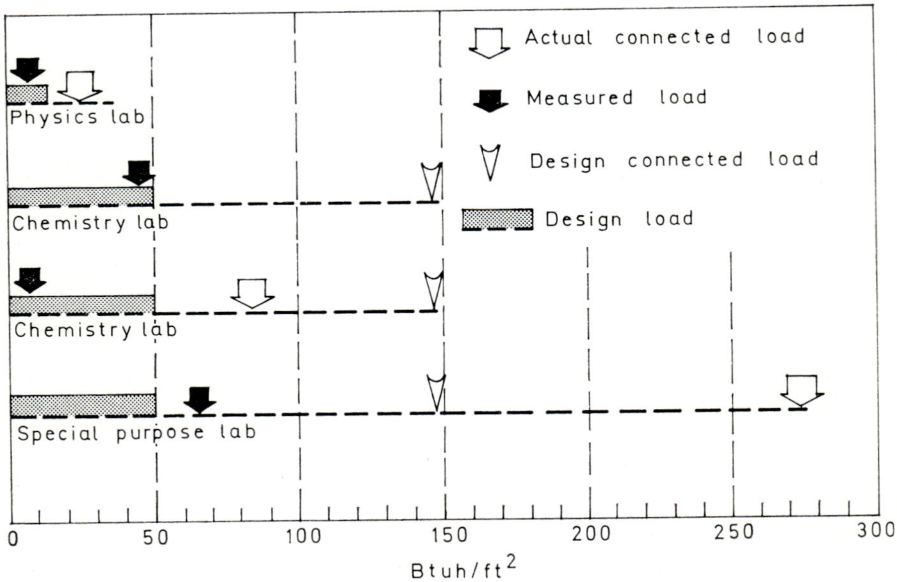

FIGURE 33 *Design and actual heat load from laboratory equipment. (By courtesy of H. E. Twietmeyer, General Motors Corporation Research Laboratories, Michigan, USA.)*

design the system for maximum flexibility to meet a varying demand in any part of the building; adequate provision should also be made for what is likely to be a rapidly increasing future requirement. In order to determine the treatment required for the supply and make-up water, an analysis of the local water supply should be obtained and compared with the purity recommended by the manufacturers of the equipment.

In the preliminary planning of the David Rivett Laboratory (*see* p. 104), provision had to be made for the dissipation of the considerable quantity of heat to be generated by equipment; this included three spectrometer electromagnets (generating from 3–8 kW each), diffusion pumps on three mass spectrometers and the pumping system of electron microscopes, X-ray generators and an electron-diffraction camera. It was estimated that 91 000 gal of water would be required each week. A cooling water system was installed; it is reticulated to 100 outlets and has been extremely satisfactory. Details are as follows:

The system operates 24 hours a day, seven days a week.

Supply and make-up water is filtered through a sand and gravel bed.

Water is cooled to 51·8–53·6°F by two 3-ton compressors; two thermostats are provided—one set to 51·8°F and the other to 57·2°F (the setting is changed over weekly). The temperature variation does not exceed 1·8°F.

The insulated storage tank has a capacity of 500 gal; to reduce filter replacement, every two months the system is flushed with a cleansing agent.

The water is reticulated by two 2-stage centrifugal pumps (with one standby) at rates of 900–1500 gal/h and at a maximum pressure of 60 lb/in^2. Each of the two floors in the three wings has a separate flow and return system which is connected to a series of bypasses and

FIGURE 34 *Cooling water outlet; CSIRO David Rivett Laboratory, Clayton, Vic.*

valves; this makes it possible to meet a demand which can vary from a load requiring one pump operating all six systems, to two pumps operating the two systems in one wing.

The flow pipe is 1 in diameter and the gravity return is $1\frac{1}{4}$ in; both are coated with Seculate anti-condensation paint. The pipes and tanks are copper; the return water passes through a strainer before it re-enters the tank.

A typical outlet is shown in Fig. 34. The pipework and a cup sink are enclosed in a PVC unit with a hinged transparent lid. Cooling water is circulated through the equipment by means of a hose connected at the left-hand side; it returns through a hose connected at the right-hand side. At the inlet connection, a small quantity of water is diverted into the cup sink where it can be seen; this water then drains back to the main return pipe. Whilst both open and closed systems can be used, the open visual method does provide a measure of safety in that any breakdown is readily detected.

If for any reason (*e.g.* power failure) the flow of cooling water is interrupted, a flow switch and solenoid valve immediately changes the supply over to the mains.

In order to obtain a pulse-free supply of cooling water to some equipment, it was necessary to fit pulse attenuators in the plant room and close to the equipment.

Rubber flexible connections were fitted to the flow pipes; these were too rigid and were replaced by Anaconda vibration eliminators which have a bellows-type seamless flexible tin–bronze core.

HOT AND COLD ROOMS

Prefabricated units are recommended because the construction and finish are superior to rooms built *in situ*; furthermore, they can easily be dismantled and re-erected elsewhere if this becomes necessary. Prefabricated construction is particularly suited to cold rooms because there is less likelihood of the vapour barrier being damaged; also, the internal finishes of aluminium or plastic laminate are preferable to plaster and paint which can be a source of constant trouble.

Some other points regarding the design of hot and cold rooms are:

For a hot room, a simple installation consisting of a 500 W heater and a small fan running continuously is generally adequate.

For a cold room 10 by 12 ft, a 1 hp unit with a 1 hp standby will maintain $36° \pm 1°F$; the standby operates only if there is more than the usual heat or traffic.

These rooms are occupied for short periods only and therefore the supply of air is usually sufficient; it is supplemented by fresh air drawn in each time the door is opened. In a cold room, it is a wise precaution to install a small exhaust fan near the ceiling at one end and inlet louvres near the floor at the opposite end. The fan can be switched on every morning to draw in several changes of fresh air; it can also be operated if the room has to be exhausted quickly in an emergency.

A note of caution regarding the use of particle board for bench tops and shelves, particularly in a hot room: in the manufacturing process, insecticides and fungicides are added, and these can contaminate the air sufficiently to cause abnormalities in some experiments.

Chapter 6

Fire and Safety Protection

In laboratory design, fire and safety protection is a basic consideration because of the commonplace presence of flammable liquids, combustible chemicals, and compressed gases, etc. Not only is the risk greater than in office-type buildings, but the cost of replacing fire-damaged scientific equipment is enormous. This hazard can be reduced to an absolute minimum by sound planning, careful selection of building materials, and good housekeeping.

Extracts from an article 'The Fire Resistance of Timber' written by R. G. Pearson in a CSIRO Forest Products Newsletter state:

> 'The burning contents of buildings may create temperatures of 1700°F and upwards. All structural materials are affected in some manner at such temperatures. Steel begins to lose strength as the temperature approaches 500°F. At 1000°F, steel has one-half, and at 1400°F only 10% of its original strength. . . .
>
> 'Significant reduction in bond between concrete and steel occurs when temperatures reach about 600°F. However, heat penetrates slowly, and reinforced or prestressed concrete with a thick covering of concrete over the steel can retain much of its strength for an extended period in a fire. . . .
>
> 'Even in intense fires, large timber sections burn at a rate of only about $1\frac{1}{2}$ in of depth of charcoal per hour. . . .
>
> 'The direct effect of heat on the strength of the material is not the only consideration. The high coefficient of expansion of steel causes steel girders and trusses to expand markedly in a fire. The resulting push outwards on their supports is often sufficient to collapse masonry walls and so bring down the entire structure. Such expansion does not occur with timber and this hazard therefore does not exist.'

Fortunately, not many people have seen a fire in a laboratory; for this reason I thought it would be of interest to describe the extent of damage caused by a recent fire. The building was in a built-up area, the fire occurred after working hours, and the fire brigade was alerted automatically by the thermal alarm system. Most of the damage was restricted to one large laboratory. The partitions were timber-framed, sheeted both sides with hardboard, and painted; above bench height, these were completely destroyed. The fibrous plaster ceiling remained intact. The bench units were damaged, but most of the bottles and equipment in the cupboards were not effected. Loose papers lying on benches were burned, but papers in drawers were not damaged. Water from the fire hoses knocked bottles off the reagent shelves, and the resultant fumes from acids and chemicals probably hampered operations more than the actual fire. One surprising sidelight was that several polythene winchesters which were partly full of acetone remained intact, but the top portion softened and sagged down to the level of the solvent. Quite apart from the cause of the fire, the lessons to be learned are:

Partitions should be of incombustible materials—preferably with a 1-hour fire rating.
Storage of chemicals in a laboratory (particularly on reagent shelves) should be reduced to a minimum.
There should be notices—prominently displayed—indicating the positions of main shut-off valves and switches for power and air-conditioning.
One or two floor drains should be provided on each storey to dispose of the minor flooding caused by fire hoses.

The extent to which fire-resisting construction must be used, the location and number of fire-isolating staircases, fire isolation of lift and service shafts, fire walls, etc., must comply with building regulations. It is standard practice for provision to be made in the plans and specifications for automatic fire detection and fire service, including hydrants. The preliminary plans are submitted to the local fire authority for approval, and this body stipulates the number and type of hand extinguishers which should be installed, and checks that the design provides an adequate and reliable water service with sufficient pressure at the highest point on the site.

FIRE RATING OF WALLS, CEILINGS AND DOORS

The Commonwealth Experimental Building Station, in an article entitled 'Fire-resisting Construction' (N.S.B. No. 87), states:

> 'Resistance to fire is graded according to the period of time, expressed in hours, that a specimen of the construction is able to withstand standardised test conditions of temperature and loading. The conditions of test are broadly similar in all countries where determinations of fire resistance are made, and for Australia they are presented as Australian Standard No. A30-1958, Section Four. Bare metal framework in particular, and light timber framing, exhibit a low order of fire resistance under the standard condition and both types of construction fail by a wide margin to qualify for a grading of 1-hour fire resistance which is the lowest grade recognised in Australian building regulations. An 11-inch cavity wall of clay brickwork, in contrast, is able to withstand the test conditions for more than six hours and thereby exceed the most stringent requirement of Australian building authorities which is a 4-hour rating.'

The fire resistance ratings of some materials are:

Walls

External and internal bearing walls, fire walls

4 hours — 11 in cavity clay brick
9 in solid clay brick
8 in solid concrete block
7 in reinforced concrete

External panel walls, internal non-load-bearing walls and partitions, lift, stair, and shaft enclosures

3 hours — 6 in reinforced concrete
2 hours — 5 in reinforced concrete
$3\frac{5}{8}$ in solid concrete block with $\frac{1}{4}$ in plaster both sides

1½ hours — 4½ in solid clay brick
1 hour — 3⅝ in solid concrete block
3⅝ in hollow concrete block with ⅜ in plaster both sides
3⅝ in light-weight hollow concrete block
Framed walls of 2½ in metal studs, sheeted both sides with ⅝ in plaster board
Framed walls of 3 in by 2 in timber studs, sheeted both sides with ⅝ in plaster board or ½ in fibrous plaster

Floors, roofs and ceilings

3 hours — 5½ in solid reinforced concrete with 1 in minimum cover to reinforcement

Doors

1, 2, and 3 hours — Metal-faced sliding doors, and timber-veneered hinged doors of approved construction and fittings
1 hour — Solid core timber door 1¾ in thick

Particular attention should be given to the fire rating of partitions in laboratories. Too often they are sheeted with combustible materials, and this surely is an unnecessary fire hazard when one considers that bottles containing flammable liquids and chemicals are kept on shelves fixed to the partitions, and there are power outlets, hot plates, and bunsen burners on the bench top. It is recommended that all partitions should be of incombustible materials, preferably with a 1-hour fire rating.

FIRE WALLS

In laboratory buildings, special consideration should be given to the provision of fire walls. Regulations require that, in a building of high hazard occupancy which is not equipped with a sprinkler system, the maximum distance from any point to an exit shall be 80 ft. It is suggested, therefore, that there should be fire walls across the full width of the building in positions so that the length of corridor between fire doors does not exceed twice this distance—*i.e.* 160 ft. Regulations do not require fire barriers in large roof spaces or service corridors; however, the fire authority in Melbourne recommends that fire barriers (of approved incombustible materials) should be provided if the roof space, penthouse, etc., exceeds an area of 10 000 ft^2. In laboratory buildings it is desirable to provide fire barriers for smaller areas. Of course, access must be provided between the sections.

FIRE BREAKS IN VERTICAL DUCTS

In the event of fire, a vertical duct very quickly becomes a chimney and a connecting link for smoke and fire to spread rapidly into the laboratories directly above. Regulations require air ducts to be fitted with fire dampers operated by fusible links, and pipe ducts to be fire-stopped at each floor with material having the same rating as the floor. In the Homer Research Laboratories described in Chapter 8, water and fire stops were provided at every hole where a pipe passed through the floor. A metal sleeve was cast into the concrete slab; it projected 1 in above the finished floor, and the space between the pipe and the sleeve was filled with weak-mix concrete.

HIGH-RISK AND OVERNIGHT LABORATORIES

When a laboratory is set aside for high-risk or overnight experiments, it follows that it should be designed to give maximum fire protection and safety. It should incorporate the following:

Walls and door—2-hour fire rating, and no large areas of glass—*e.g.* windows of minimum size, and no highlights.
Ceiling—3-hour fire rating
Floor—Concrete with floor waste and 3 in high kerb across door openings.
Fittings—Benches to have metal frames and incombustible top (*e.g.* compressed asbestos cement), and to be open underneath; fume cupboard bench tops and linings to be acid-proof tiles or compressed asbestos cement; wall cupboards to be steel.
Power and light fittings to be flameproof.
Automatic fire protection system.
Fire and safety equipment, and isolating valves to all services; these to be readily accessible in the corridor.
Where there is risk of explosion, a blow-out panel should be provided.

FIREPROOF ROOMS

At least one fireproof room should be provided in every laboratory building in order to give maximum protection to valuable items of equipment and irreplaceable records. Properly used, a fireproof room is an essential adjunct to every laboratory. Although it need only be quite small, it is usually one of the first rooms to be deleted in the planning stage—because of shortage of space or funds.

FIRE-RESISTANT CABINETS

Unfortunately, problems often arise when even a fireproof room is shared by a number of people. An alternative which offers a satisfactory solution is to provide individual fire-resistant cabinets in laboratory offices. A suitable unit on the market measures 13 in high by 18 in wide by 24 in deep externally; it is double-walled steel with 1 in thick asbestos packing, and it is fitted with one adjustable shelf and a drop-front door with lock.

STORAGE OF ACIDS, TOXIC AND HAZARDOUS CHEMICALS AND SOLVENTS

If it is necessary to have large quantities of acid in a laboratory, the bottles should be kept in a special cupboard when not in actual use. The cupboard should be lined with asbestos or PVC sheet, and be fitted with a tray. Generally it is located underneath a fume cupboard where it has the advantage that any fumes are exhausted—but the disadvantage that it is directly beneath the most likely trouble spot in the laboratory.

Toxic and hazardous chemicals in small quantities are kept on a shelf within the fume cupboard, or in a ventilated cupboard underneath it.

The quantity of solvents housed in a laboratory can vary from one to 30 litres. Again, if there are large quantities, the bottles should be kept in a cabinet when not in use. Suitable cabinets are available, the specification for them being: 16-gauge steel double walls and door with $1\frac{1}{2}$ in air space, vented, adjustable shelves, self-closing doors with vapour seal and spark-free catches, and a sump having a capacity of not less than 25% of the maximum contents of the cupboard and not less than that of the largest bottle.

In some laboratories (*e.g.* organic chemistry) even larger quantities must be readily available. To meet this situation, one or more stores should be provided on each floor; these are, in fact, intermediate between bulk store and laboratory. The walls and door should have a 2-hour fire rating, the floor should be recessed with duckboards flush with the main floor, and there should be an exhaust system operating continuously.

FLAMMABLE LIQUIDS STORE

Bulk supplies of flammable liquids should be stored in a separate building with maximum isolation from other buildings; there should be road access and a ramp. The Fire Underwriters' Association of Victoria requires that the building be constructed as follows:

Walls—9 in brick or 6 in reinforced concrete
Roof—4 in reinforced concrete
Floor—Concrete
Opening—To be protected by single automatic tin-clad fire door
Sill to opening—To be raised to form a sump capable of holding at least 25% of the total quantity of liquid stored, but in no case to be less than 3 in. in height.
Ventilation—To be direct to atmosphere through steel vents at sill and ceiling height; vents to be fitted with 28-mesh wire gauze and to have an area of not less than 100 in^2 per 1000 $ft.^3$ If the store is not a separate building, the ventilation ducts must be masonry—extending through the main risk to open air.
Electric lights and fittings—Suitable for hazardous locations.
Fixtures—To be of incombustible materials, except that shelves can be timber.

Sometimes a hinged monorail with hand-operated hoist is provided to lift 40-gallon drums into racks; the monorail should be earthed for static electricity. Spring-loaded brass or stainless steel ball valves with teflon packing have given satisfactory 'no-drip' service when fitted to drums.

MULTIPLE EXITS

It is wise for all laboratories to have an alternative exit; all large and special laboratories, and those fitted with fume cupboards, certainly should have one. This exit can be through an instrument laboratory or office between two laboratories, or a narrow emergency door between adjoining laboratories (*see* Fig. 55). In some situations, a hinged or knock-out panel above bench level will be adequate.

POSITION OF POWER OUTLETS

Power outlets over benches should be mounted on the rail under the reagent shelf, as shown in Fig. 7. In the CSIRO standard detail, the centre of the outlet is 13½ in above the bench; this conforms to Australian Standard CCI Part 1—1969, Wiring Rule 6.23, which states that, in laboratories where flammable solvents are used, power outlets shall be installed at a height of not less than 4 ft above floor level and not less than 12 in above a bench. This Rule also states that the interior of a fume cupboard in which volatile liquids or gases are used shall be considered to be Class 1 Division 1 location—*i.e.*, flameproof lights and power outlets shall be used. Standard power outlets shall not be used within an area of 24 in. in all directions except vertically above the top of the face opening of the cupboard.

Admittedly, it is necessary to use flameproof power outlets in certain circumstances (*see* p. 90); but, for most fume cupboards, I consider that it would be most unrealistic to install them. The fittings are, of necessity, bulky and cumbersome, and they would discourage rather than encourage the laboratory workers to use the fume cupboard; furthermore, as most of the equipment within the cupboard is not flameproofed, they could lure the operator into a sense of false security. Another point is the greatly increased cost; the 1971 price of a 20 A 250 V flameproof switch-and-plug combination and plug top is $88.

The Wiring Rule means that the standard power outlets can only be used for a fume cupboard if they are fitted in the following positions:

On the rail across the top of the face opening (in which case the cables dangle dangerously in front of the sash).

At the side of the cupboard above the level of the face opening.

On the wall 1 ft above the bench and 2 ft from the back corner of the cupboard. (In this case, the cable must either pass through a neoprene iris in the side of the cupboard, or trail across the bench and back into the cupboard.)

There are many scientists who are very unhappy with this regulation. It seems that an attempt has been made to determine the safest position for the power outlet, but little thought has been given to the person who has to reach it, get the flex into the cupboard, operate the sash, and use equipment with built-in switches which certainly are not flameproof. I hope this regulation will be amended to allow standard power outlets to be mounted on the mullions at each side of the fume cupboard at a height of not less than 1 ft above the bench; the cable then can pass through the face opening (without danger from the sash), or it can pass through a neoprene iris in the mullion just above the bench top.

AUTOMATIC FIRE DETECTION SYSTEMS

All laboratories should have automatic fire detection equipment which simultaneously sounds an alarm in the building and at the fire station. Systems available at present include thermal alarm, sprinkler, smoke detector, flame detector, carbon dioxide, and halogenated hydrocarbon. In built-up areas where the services of a fire brigade are readily available, it is usual to install thermal alarms in buildings of medium risk, and sprinklers in buildings of high risk; the other systems are designed for special conditions.

Thermal alarm systems are operated by detectors in the ceiling. The maximum permissible area per detector is 500 ft^2—provided the ceiling is flat and the floor-to-ceiling height does not exceed 10 ft; another requirement is that the detectors should not be more than 10 ft from any wall. The cost of thermal alarms is less than one-third that of sprinklers. It is fire brigade policy to arrive on the site within four minutes of any call, and many cases have been recorded when it has been necessary for them to use only hand extinguishers—thus eliminating or reducing damage by water; in laboratories, this is an important but not overriding consideration.

Sprinkler systems consist of water outlets installed in the ceiling—one every 100 ft^2. The outlets are sealed by a heat-sensitive device which can be set to operate at a temperature within the range of 155–165°F. At a pressure of 20 lb/in^2, the water discharged from a $\frac{1}{2}$ in head is 20 gal/min, and it will continue until the main supply valve is shut off. This type of installation attacks the fire before it has time to get out of hand, and it requires no human assistance; its efficiency is not impaired by the presence of smoke or fumes, nor is it hindered by difficulties of access. One firm which has been operating since 1886 claims that, of the 4500 fires which have occurred in buildings protected by their sprinklers, successful 'saves' have been 99·7%. Of these, 67% were extinguished by the operation of only one sprinkler, and 83% by two sprinklers.

The principle of smoke detection is based on an element which is sensitive to the products of combustion, whether visible or invisible. Because its operation is not dependent on any heat being generated, it gives the alarm much more quickly than thermal alarm or sprinkler systems; when used to actuate a carbon dioxide gas system, the fire is extinguished before any major damage is done. One detector will protect an area of 1000 ft^2. This type of system costs more than thermal alarms, but its extreme sensitivity makes it ideal for the protection of the more expensive types of scientific equipment.

Carbon dioxide gas is a very efficient fire extinguishing agent. It is dry, does no damage, is non-corrosive, and is a non-conductor of electricity. A built-in automatic carbon dioxide system is particularly suitable for a high-risk situation; it gives immediate detection and rapid extinguishing. The newer BCF (bromochlorofluoromethane) is even more effective, but at present it is more expensive. The following is a description of a CO_2 installation in a high-risk fume cupboard which only approved staff are permitted to use:

> The cupboard is 12 ft long and is used for 3000 V paper electrophoresis under toluene. The four glass tanks contain a total of approximately 40 gallons of toluene and are supported on a timber slatted platform over a stainless steel sink which extends the full length of the cupboard. The waste pipe and trap are copper.
>
> A stainless steel standing waste is used to prevent a large volume of solvent being discharged into the waste if a tank is broken. This standing waste has two rows of holes near the bottom to drain the small but constant flow of cooling water from the tanks.
>
> When the sash is lowered, a red light shows and a timing clock starts; then power is turned on from a power pack outside the fume cupboard. When the sash is lifted (even inadvertently), it operates a trip switch which disconnects all power outlets.
>
> Fireproof screens and an automatic CO_2 system are installed. The screens are timber-framed,

sheeted both sides with 20-gauge stainless steel, and filled with 1 in thick magnesia asbestos blanket. They are released automatically by a CO_2 operated mechanism, but a time lag (sufficient for the scientist to get his head out of the way) has been provided by a perforated piston in an oil-filled cylinder. When actuated, a detector automatically releases CO_2 from one 20 lb cylinder; this is discharged through two nozzles and operates two micro-switches—one to disconnect all power, and the other to turn off the exhaust fan. A duplicate system is provided—also operated by one 20 lb cylinder and with external manual controls.

An automatic water spray system is recommended for the protection of rooms where large quantities of flammable liquids are stored. This is a special application of a sprinkler system; the capacity of the heads is greater and they are closer together. It is more effective than CO_2 gas which is prevented from acting as a blanket because of the strong up-currents of air from the fire.

HAND EXTINGUISHERS, HOSE REELS AND HYDRANTS

In addition to an automatic fire detection system, provision must be made for the installation of hand-operated fire extinguishers and hydrants. Instructions regarding the type and location of these fittings are given by the fire authority.

Hand extinguishers should be installed in the corridor, close to the escape stair, on the basis of one to every 2000 ft^2. In addition, one extinguisher should be kept within every laboratory, near the escape door.

There are several types of extinguishers. It is usual to install soda-acid or high-pressure water in the corridor for general use, and CO_2 gas in the laboratories to cope with small flammable liquid or electrical fires. In addition, Tri-class dry chemical extinguishers are available for class A (paper and wood), class B (flammable liquids), and class C (electrical) fires, foam for class B, dry chemical (sodium or potassium bicarbonate and pressurised nitrogen) and halogenated hydrocarbon for class B and C fires. On the basis of equivalent test fires as specified by the Standards Association of Australia, the following weights of extinguishing media are required: 2 gal foam, 5 lb dry chemical, 5 lb halogenated hydrocarbon, and 10 lb carbon dioxide.

SAFETY SHOWERS

All laboratories with fume cupboards, and some special laboratories should be fitted with a safety shower. So far as possible, the showers should be in the same relative position in each room so that, in an emergency, they are easily found. CSIRO practice is to have a safety shower over the sink at the end of the peninsular bench on the fume cupboard side of the laboratory, as shown in Figs. 1 and 7; in this position, any drips are caught. The shower fitting has a quick-acting valve which is operated by a red polythene-covered lever. In many overseas laboratories, showers are provided over every door to the corridor, as shown in Fig. 53 (it is not usual to have a floor waste), and an eye bath is fitted on one of the benches.

ADDITIONAL PRECAUTIONS

There are many more fire and safety precautions which could be mentioned—some of them seemingly trivial; but no item should be dismissed if it contributes to maximum safety. A few are:

Fire-fighting and first-aid equipment should be located close to the escape stairway and should include medical equipment, fire blanket, breathing apparatus, and sand bucket.
There should be adequate ventilation and fume exhaust systems.
Bottled gases (including burner gas) should be reticulated rather than cylinders being manhandled in a number of laboratories.
The number and positions of service outlets (especially power) should be such that double adapters and long lengths of tubing and flex are not required.
Anti-static flooring should be installed in hazardous locations.
Special provision should be made for the disposal of hazardous waste.
Waste bins should be metal, with self-closing lids.
Where 'No Smoking' is applicable, the signs should appear not only on one side of the doors, but both sides.

No matter how many fire and safety precautions are provided in a building, they cannot be really effective unless the staff is safety conscious. Built-in fire protection is provided at considerable cost, but it can be largely nullified if, for example, excessive quantities of solvents are kept in the laboratories, or safety containers are not used to carry winchesters of acids or solvents from the store—and so on! The following are extracts from an article 'Fire and Explosion Problems in Laboratories and Pilot Plants' by M. M. Braidech, American Insurance Association, New York, which was published in the Journal of Chemical Education, April 1967:

'The prime object of any organised safety program is to stimulate an awareness for hazards and to convince the persons concerned that precautions are necessary and that they should be observed. . . . Experience with successful programs has shown that while 'top-man' interest is important, the administration should participate only as a guiding hand and allow the work of safety to come from a laboratory committee and the immediate supervising staff. Based on this viewpoint the following procedure for safety effort is recommended:

'Organise a laboratory safety committee among principal staff members to maintain sustained and vigilant interest on a rotational basis.
'Conduct a self-survey on fire hazard and explosion potentials and evaluate their degree of risk. Review every phase of the laboratory program for faulty techniques. Consider storage, handling, and residue or waste disposal problems. Inventory (use and location) all hazardous chemicals and determine their general flow into the various laboratory units of the building.
'Formulate a suitable safety code. . . . Consider segregation and quantity limitation of hazardous chemicals on laboratory benches and reagent shelves, laboratory supply rooms, and main stock rooms. Instruct custodial personnel on proper handling and storage procedures.

'Maintain an active safety interest—

Proper training should be given in inspection and hazard spotting, emergency procedures in event of fire, and proper use of fire extinguishing equipment. Organise a floor-to-floor mutual aid program and schedule drills.

Safety indoctrination of new personnel should be an established policy and an assigned committee function.

Periodic spot inspections (preferably unannounced) should be made by the committee to determine the safety conditions under actual routine operations.

Safety meetings should be planned and scheduled. Draw attention to any 'near-miss' experiences in the laboratory and to accidents reported in technical publications. Keep the staff posted on new texts and manufacturers' bulletins pertinent to chemical safety. Temper enthusiasm with achievement. Invite outside speakers on safety.'

If you have not already done so, I suggest that you read the excellent series of articles entitled 'Safety in the Chemical Laboratory', which appeared in the Journal of Chemical Education during the period January 1964 to January 1967. These now are available in book form under the same title and edited by N. V. Steere.

Chapter 7

CSIRO Laboratories

The research programme of CSIRO is carried out by thirty-four Divisions and associated Field Stations which are established on eighty-five sites throughout Australia. The value of major buildings costing over $50 000 which were approved for construction during the five-year period ending June 1970 averaged $3 million per annum.

This chapter describes nine laboratories which, although varying considerably in size, are representative of the type of accommodation provided for the wide range of research activities. All these laboratories incorporate modular planning and standard fittings, and—wherever possible—repetition of standard layouts.

HORTICULTURAL RESEARCH LABORATORY, GLEN OSMOND, SA

This quite small laboratory is a good illustration of the wisdom of providing adequate space for expansion. The first stage of the two-storey building was 70 by 48 ft, with the main entrance at

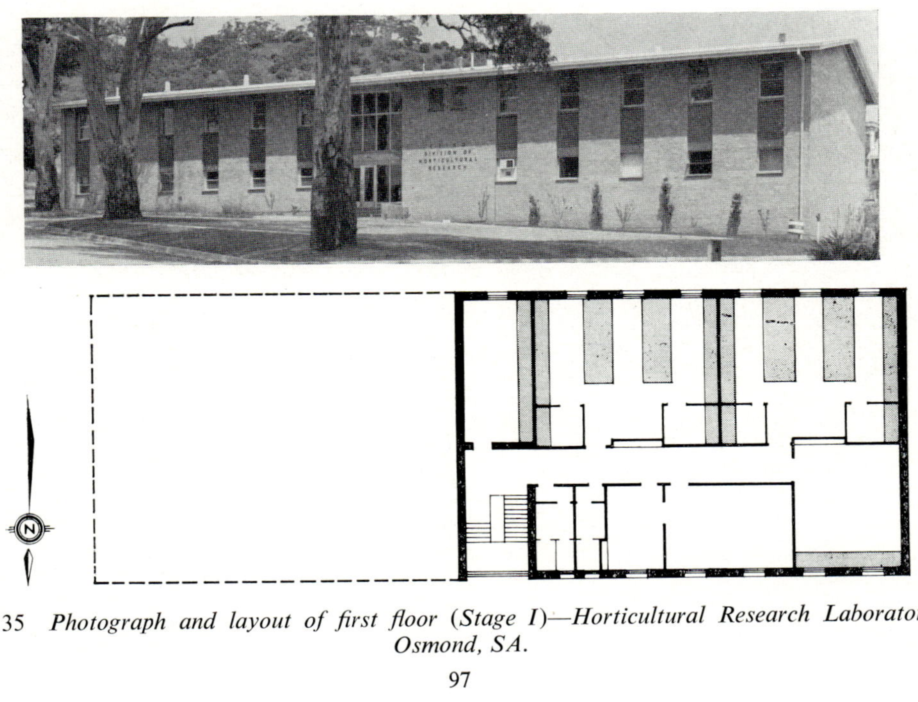

FIGURE 35 *Photograph and layout of first floor (Stage I)—Horticultural Research Laboratory, Glen Osmond, SA.*

one end and provision for doubling the length. On the first floor there are administrative offices, two 30 by 24 ft laboratories, and an instrument laboratory; the ground floor was left without partitions to allow for future expansion. This section of the building cost \$70 000 and was completed in 1964 when the total staff was ten; two years later the ground floor had been partitioned and was fully occupied. In another two years when the staff was 23, the 60 ft extension was completed at a cost of \$60 000; again, most of the ground floor was left available for expansion. (*See* Fig. 35.)

BIOCHEMISTRY LABORATORY, DIVISION OF PLANT INDUSTRY, CANBERRA, ACT

The Biochemistry Laboratory was built in 1961 at a cost of approximately \$15 per square foot. It is a three-storey building 152 by 47 ft—brick and reinforced concrete with a roof space. The windows are north- and south-facing; glazing is continuous on the south elevation, and on the north there are 6 ft wide windows with hoods over. The building is heated by hot-water radiators.

The module is 10 ft, and the floor-to-ceiling height is 9 ft. The 5 ft 6 in wide off-centre corridor has seven 30 by 24 ft and two 20 by 24 ft laboratories of standard layout on the south side, and service laboratories 10 and 20 ft by 14 ft on the north side. Two 8 by 6 ft laboratory offices are provided within each three-module laboratory.

The standard peninsular bench is 14 by 5 ft; it has a writing space 3 ft 6 in wide at the window end on one side, and the following service outlets:

4 cold water over each of 2 troughs
2 cold, and 1 hot and cold water, over sink at end of bench
4 power and 1 gas on each side

Fume cupboard exhaust fans, etc., are installed in the roof space and have a vertical discharge. The roof is corrugated asbestos cement. (*See* Fig. 36.)

LAND RESEARCH LABORATORY, CANBERRA, ACT

The first stage of this building was completed in 1964; it is a one- and two-storey building with laboratories, administrative offices, library, conference room, and canteen. The second stage was built in 1968; it is a three-storey building 220 by 47 ft, with laboratories and space for photographic interpretation, etc., on two floors; the ground floor has been left open for expansion. The total area is 58 000 ft^2, and the cost was approximately \$13 per square foot.

The building is brick with reinforced concrete frame; the columns and hoods over the windows are precast exposed aggregate. Fume cupboard exhaust fans are installed in the roof space, and they have a vertical discharge. The roof is corrugated asbestos cement.

The module is 10 ft and the floor-to-ceiling height is 9 ft. The 5 ft 6 in wide off-centre corridor has 24 ft deep laboratories on the south side, and 14 ft deep laboratory offices on the north side. (*See* Fig. 37(b).)

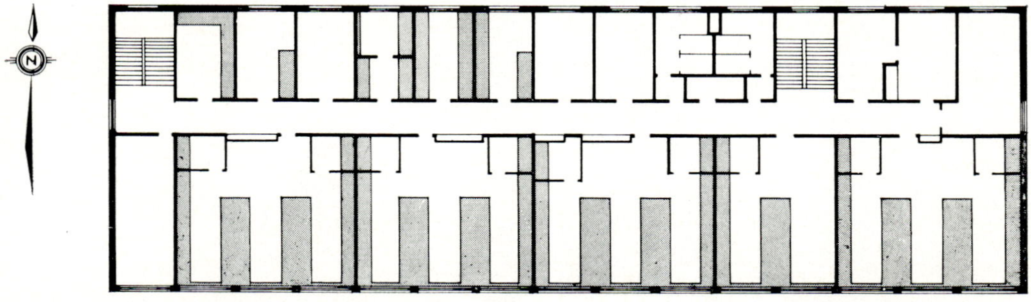

FIGURE 36 *Photograph and layout of first floor—Biochemistry Laboratory, Division of Plant Industry, Canberra, ACT.*

FIGURE 37(a) *Perspective sketch—Land Research Laboratory, Canberra, ACT.*

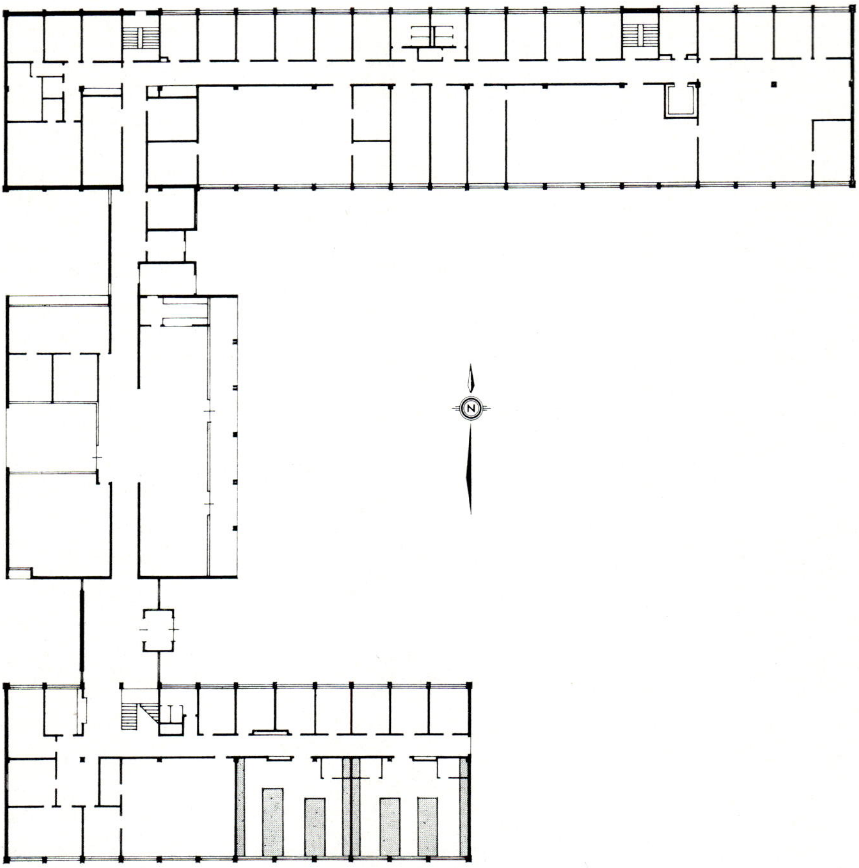

FIGURE 37(b) *Layout of ground floor—Land Research Laboratory, Canberra, ACT.*

SOILS LABORATORY, CANBERRA, ACT

This three-storey brick and reinforced concrete building has an attached four-storey 42 by 22 ft rain simulator tower. The total area is 24 100 ft^2, and it was completed in 1971 at a cost of approximately $22 per square foot. (*See* Fig. 38.)

The module is 10 ft, and the floor-to-ceiling height is 9 ft. The 5 ft 6 in wide off-centre corridor has 27 ft deep laboratories on the south side, and 12 ft deep offices on the north side; some of the laboratories have not been fitted with benches. The partitions are Besser block, unplastered, and the building is heated by hot-water radiators.

The standard peninsular bench is 21 by 5 ft; it has writing spaces 3 ft 6 in wide at the window end on both sides of the bench, and the following service outlets:

> 4 cold water over a trough; 1 cold, 1 distilled, and 1 hot and cold water, over sink at end of bench; 2 single air; 2 single gas; 8 power on each side.

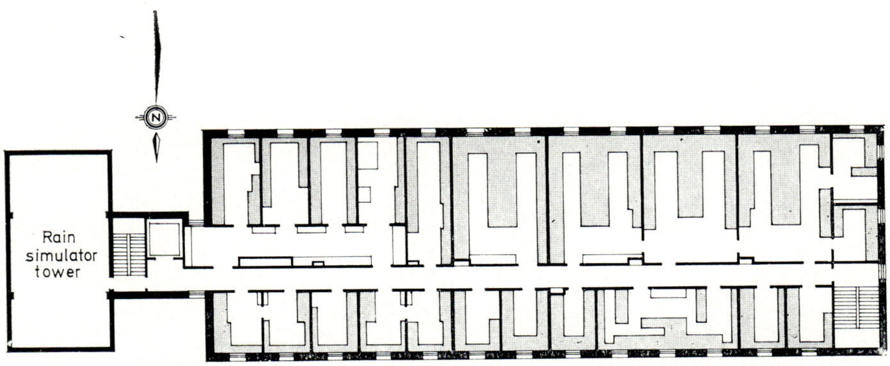

FIGURE 38 *Photograph and typical floor plan—Soils Laboratory, Canberra, ACT.*

F. C. PYE WILDLIFE RESEARCH LABORATORY, DARWIN, NT

This small laboratory was designed for tropical conditions; it is a single-storey building 133 by 42 ft with a 5 ft wide verandah on all sides. It was completed in 1971 at a cost of approximately $40 per square foot. (*See* Fig. 39.)

The external and internal walls are grey cement brick, unplastered, and the floor is vinyl tile on concrete. The low-pitched roof is steel decking supported on steel trusses, with a vapour barrier underneath and mineral-wool insulation 4 in thick supported on galvanised wire mesh. There is no gutter; the stormwater is collected in a channel formed in the concrete paving which surrounds the building. The module is 9 ft 4½ in; the depth of the laboratories (including an open corridor) is 26 ft, and the laboratory offices are 14 by 9 ft. The building is air-conditioned with packaged units in three zones.

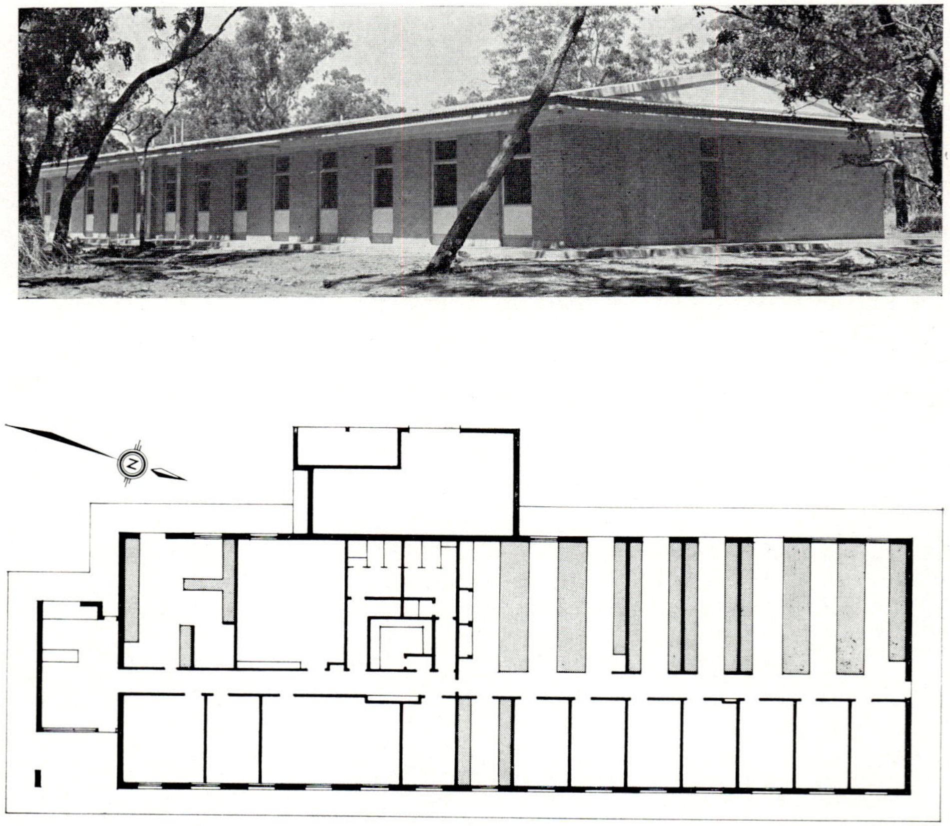

FIGURE 39 *Photograph and floor plan—F. C. Pye Wildlife Research Laboratory, Darwin, NT.*

The building site was termite-proofed. The treatment consisted of poisoning the ground below all concrete floor slabs, the foundation trenches, and an area 2 ft 6 in wide around the perimeter. After pouring the foundations, each 3 in layer of fill was consolidated and poisoned.

In the tropics, provision generally is made for a solar hot-water unit, and for all the steel (structural, door frames, etc.) to be hot-dip galvanised.

THE MAIN LABORATORY OF THE DIVISION OF PROTEIN CHEMISTRY, PARKVILLE, VIC

This five-storey building 64 by 84 ft is an extension to an existing two-storey laboratory, and it is on a restricted site. The gross area is 31 100 ft^2, and it was completed in 1963 at a cost of \$680 000 —*i.e.* \$22 per square foot. (*See* Fig. 40.)

Building details

Reinforced concrete frame with flat slab, designed for 100 lb/ft^2 on the ground floor and 70 lb on other floors. The column spacing is 20 ft to give a 10 ft module, and the floor-to-ceiling height is 10 ft 8 in. Curtain walls on the north and south elevations, with hinged panels under the windows to provide access from the balcony to service pipes; brick on east and west elevations. Internal walls are brick, plastered, or fibrous plaster on timber framing; the latter, for laboratory offices, are insulated below sill height and double-glazed above. The underside of the concrete slab is exposed; in the laboratories it is painted, and in the laboratory offices it has acoustic tiles. The floor is flexible vinyl sheet with welded joints, and coved to walls and the base of benches.

Laboratory fittings

CSIRO standard design; the bench units have a clear finish, and the tops are $1\frac{1}{4}$ in thick Queensland kauri with black aniline finish (*see* Appendix 3). The sinks are 18-gauge No. 316 stainless steel, satin finish, with polythene traps and PVC wastes. There are 26 fume cupboards, eight of which are fitted with chromatography cabinets. The standard peninsular bench is 13 by 5 ft and has the following service outlets:

5 cold water over each of 3 troughs
2 two-way gas each side
1 single compressed air each side
6 power (8 per 20 A circuit) on each side

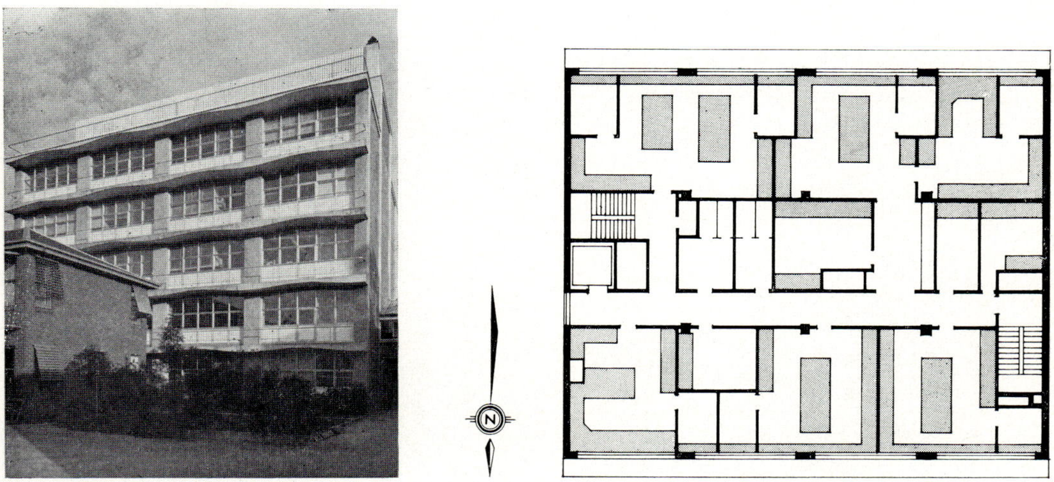

FIGURE 40 *Photograph and typical floor plan—The main laboratory of the Division of Protein Chemistry, Parkville, Vic.*

Services

Two oil-fired boilers each 875 000 Btu/hr for hot-water radiators in laboratories, heated-air system to library, dining, and conference rooms, and air-conditioning to internal service rooms. Calorifier for domestic hot water. Compressed air—Broomwade NC5 carbon ring 10 hp $7\frac{1}{2}$ ft^3 receiver at 100 lb/in^2 with oil and water filter and pressure-reduction valve to 60 lb; outlets fitted with Norgren filter regulator Series 73AC. The air compressor is on the ground floor, but all other mechanical equipment is in the penthouse. The water is treated in Cusilman bronze receivers with two 500 gal copper storage tanks in the roof, and a 5000 gal underground steel tank (with bonded lining) with booster pumps to maintain pressure of 65 lb/in^2 on the top floor. Gravity-feed water is available after office hours. Deionised water is reticulated in PVC pipe to one outlet (polythene diaphragm valve) on each floor, with provision for future extensions. The unit is Deminrolib Plant MB-9 in mixed bed column, 250 gal PVC main feed tank. Reticulated services are hot and cold water, heating gas, compressed air, and restricted reticulation of deionised water and oxygen. Copper ring mains to hot and cold water and compressed air. Cold water pipes to bench outlets are screwing brass looped to form secondary ring mains; pipe supports are fabricated steel. Electrical main switch—800 A 415 V three-phase. The central neutraliser is lined with acid-proof bricks. A Kent Mark II pH indicator-recorder-controller is installed with a Mark XX three-term pneumatic modulating control valve to feed soda-ash into the neutraliser. The pH of the effluent from the building varies from 3·7 to 5·7, and from the neutraliser it is 6·1 to 6·3.

Fire protection

Thermal fire detectors throughout, two hydrants per floor, and fire extinguishers.

DAVID RIVETT LABORATORY, DIVISION OF CHEMICAL PHYSICS, CLAYTON, VIC

The Division of Industrial Chemistry was established in 1940 on a 7-acre site at Fishermen's Bend, Melbourne. During the next 15 years the Division expanded so rapidly that it became six separate Divisions, and it was obvious that the site was inadequate. A 38-acre site adjoining the Monash University at Clayton was purchased, and Chemical Physics was the first of the Divisions to move to the new location.

FIGURE 41 *David Rivett Laboratory, Division of Chemical Physics, Clayton, Vic.*

The building has a north–south aspect and consists of four wings, with an internal courtyard; provision is made for another wing. The building has a gross area of 66 900 ft^2 and it was completed in 1965 at a cost of \$1·1 million—*i.e.* \$17 per square foot; it was designed for a total staff of 120.

A particular requirement was that the building be vibration-free, and every effort was made to achieve this. A factory approximately 330 ft from the boundary operated several forge hammers, the largest weighing 4 tons. Arrangements were made for the Bureau of Mineral Resources to test the site vibration. The results indicated that the ground displacement decreased rapidly as the distance from the hammers increased. In the first 150 ft the decrease was from about 30 units to 3 units (1 unit = 0·0001 in peak-to-peak displacement); the displacement beyond 310 ft from the 4-ton hammer was less than 1 unit. Frequencies were 12–17 cycles per second. CSIRO National Standards Laboratory commented on these results as follows:

> 'These are in general agreement with results of tests on other sites—*e.g.* a 6-ton hammer caused a maximum ground displacement of 22 units at 80 ft and 2 units at 400 ft, and a 30-cwt hammer caused maximum ground displacement of 23 units at 40 ft and less than 1 unit beyond 150 ft.
>
> 'It is very likely that other sources of vibration, closer than the hammers and perhaps within the proposed new buildings themselves, could set up vibration no less objectionable than that from the hammers. Such sources would include road traffic, air-conditioning fans, refrigeration compressors, pumps, electric motors and generators, lifts, and workshop activities.
>
> 'In addition, there is always disturbance from personnel activities. The vibration of an upper floor of a modern steel and concrete structure, due to a person walking about the floor, is quite likely to be several units, and hence could be as severe as that from the remote hammers.
>
> 'The Standards Laboratory is load-bearing construction; with normal activities in progress throughout the building, the vibration in the basement is in the order of 0·1 unit, and on upper floors 1 to 2 units have been observed due to personnel activities.'

The Chemical Physics Laboratory is two- and three-storey with a basement under one wing, load-bearing construction, reinforced concrete ribbed floors designed for 100 lb/ft^2 in the laboratories, 150 lb in the library, 60 lb in the offices, and 80 lb/ft^2 in the corridors. Each 15 ft module has three 2 ft 9 in wide fixed windows; floor-to-ceiling height is 10 ft. Partitions to laboratory offices and subdivisions in laboratories are demountable particle board panels with painted hardboard finish.

The building is air-conditioned; there are 30 package units installed to give individual control to groups of laboratories and offices; they are air cooled, reverse cycle, with split condensing units and evaporator blowers. For the rooms shown in Fig. 42, Laboratories 1–4 have one 8-ton unit, Laboratories 5–7 have a 5-ton unit, and the eight offices a $3\frac{1}{2}$-ton unit. Supply air ducts are above the false ceilings in the corridors and the air returns along the corridors to the plant room; there are 6–8 changes per hour with 80% recirculation.

Services

Reticulated services are hot and cold water, heating gas, cooling water, power and isolated earth facilities. All water passes through a sand gravel filter which has a capacity of 500 gal/hr. One 1000 kVA 22 000/415/240 V transformer is housed in a separate substation building; space is provided to double the capacity. Details of the reticulation and sub-boards in the laboratories are shown in Fig. 42. The fluorescent lights are fitted with radio frequency screens.

Fire protection

Thermal alarms and smoke detectors throughout; hydrants and fire extinguishers on each floor.

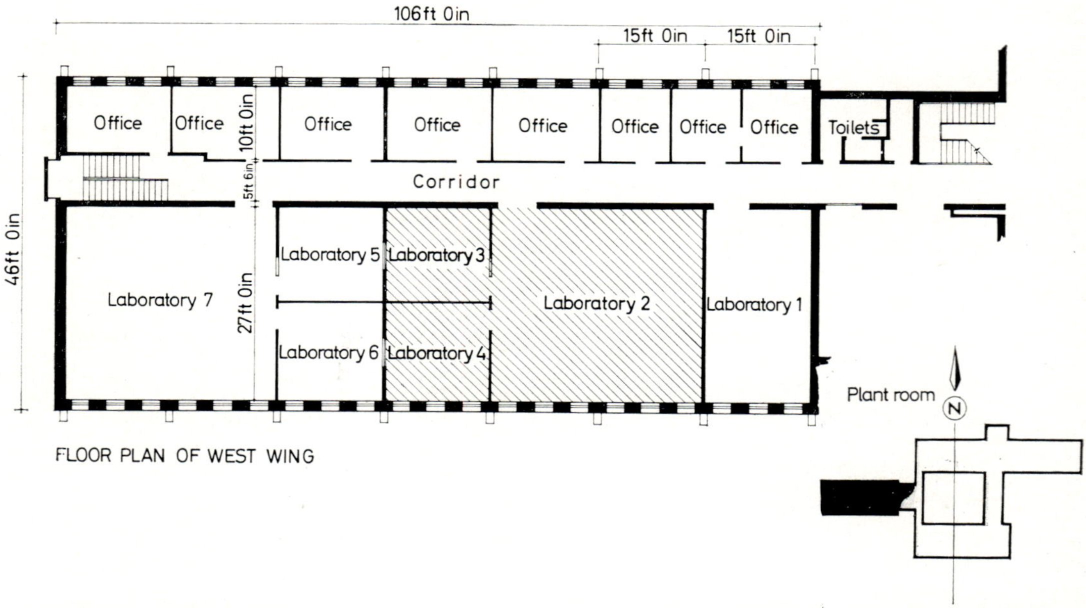

FLOOR PLAN OF WEST WING

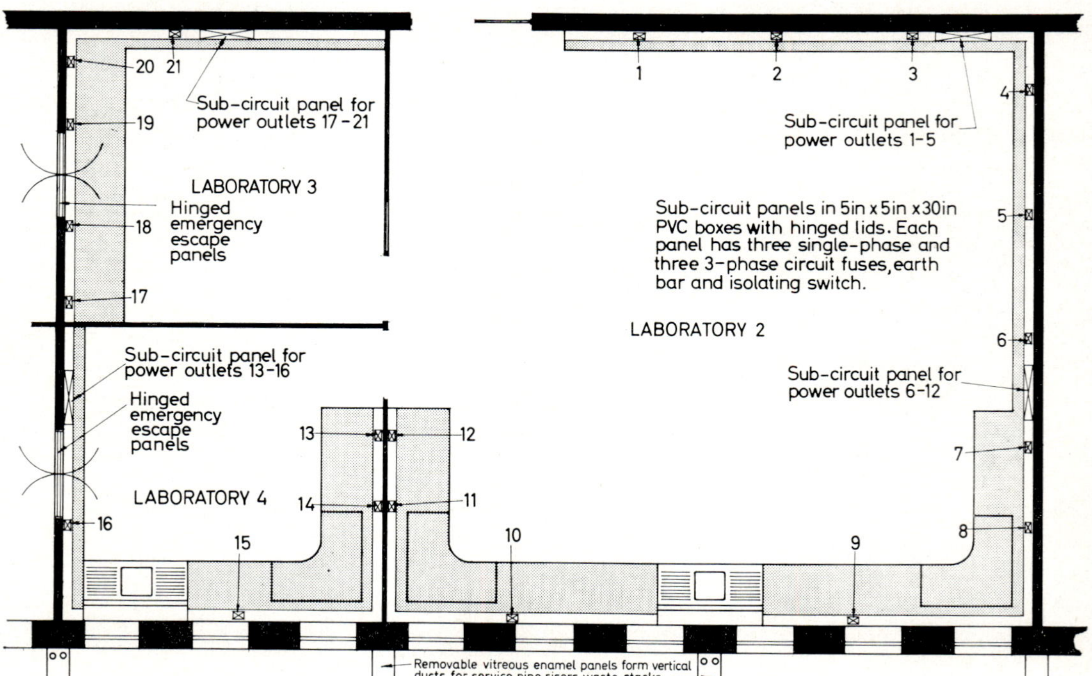

A TYPICAL LABORATORY LAYOUT

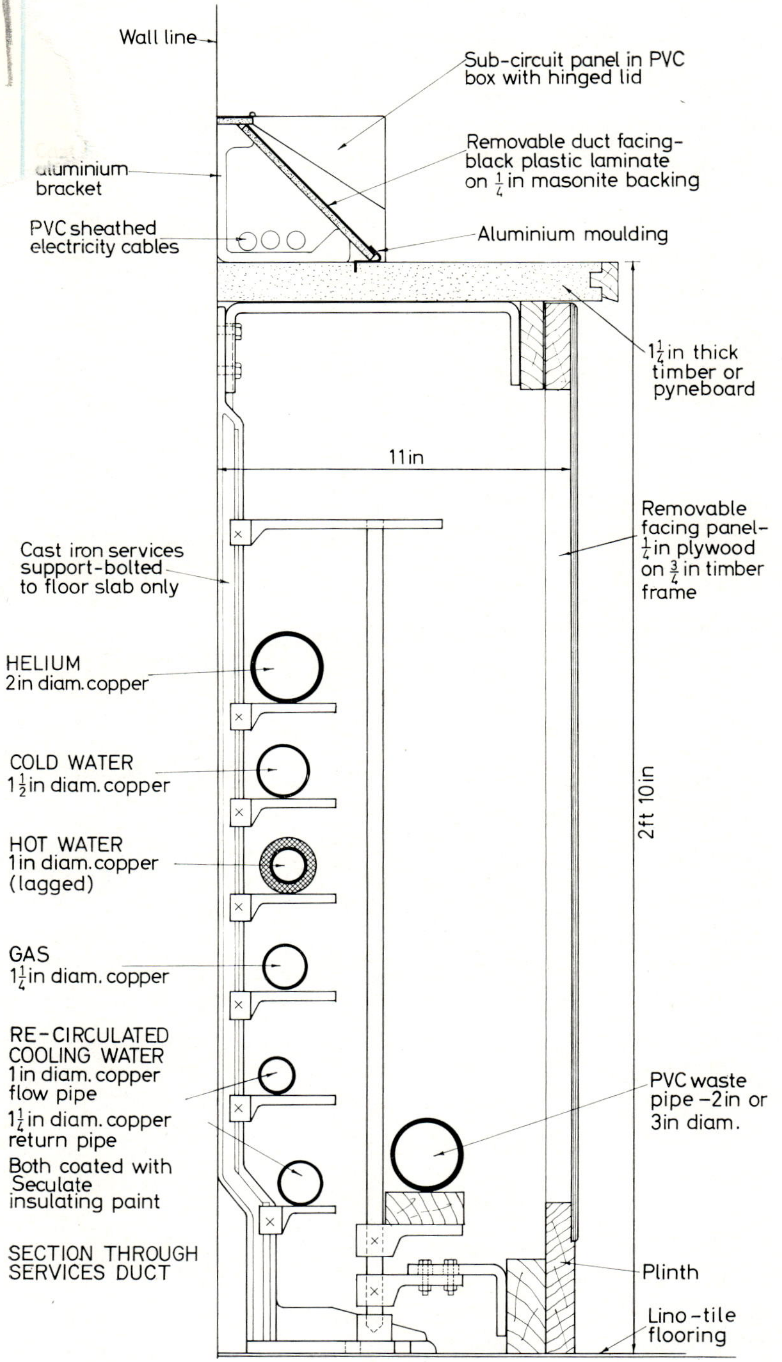

FIGURE 42 *David Rivett Laboratory, Clayton, Vic.*

CHEMICAL ENGINEERING LABORATORY, CLAYTON, VIC

This is a group of buildings consisting of a two-storey administration/laboratory building, light technical and heavy technical laboratories, process bay, workshop, plant and stores building, and flammable materials store. The buildings have a north–south aspect and have been sited so that each can be extended; they were completed in 1969 at a cost of $1·26 million, and were designed for a total staff of 100; the gross area is 58 400 ft.^{2} (*See* Fig. 43.)

Building details

The construction generally is reinforced concrete, structural steel, and brick.

Main building

Load-bearing brick, reinforced concrete, and structural steel; floors designed for 60/lb sq ft; module 10 ft; floor-to-ceiling height 9 ft.

Ground Floor	—	Off-centre corridor 5 ft 6 in, laboratories 24 ft, offices 12 ft deep.
First Floor	—	Central corridor 5 ft 6 in, laboratories 18 ft deep on both sides.

Internal walls are brick, painted; the ceiling is fibrous plaster and the floor is flexible vinyl tiles.

Light technical laboratory

Off-centre corridor 6 ft 6 in, with laboratories 19 ft and 25 ft deep; module 10 ft; floor-to-ceiling height 14 ft.

Heavy technical laboratory

102 by 51 ft; 2-ton manually operated travelling crane with 30 ft clearance below hook.

Process bay

77 by 51 ft; 2-ton manually operated travelling crane with 20 ft clearance below hook. Mezzanine floors are provided at two levels; the steel frame has bolted joints with removable gridmesh floor panels. Four 1700 ft^3/min exhaust fans with flame-proof motors are provided just above floor level. Steel columns designed to support framed experimental rigs on the paved area adjoining the building; removable wall panels to facilitate extension of the services to these rigs. The paved area has dwarf walls and is drained to an underground tank so that, in the event of large quantities of solvent being spilled, it will not enter the stormwater drains.

Laboratory fittings

CSIRO standard design; the bench units have a clear finish and the tops are $1\frac{1}{4}$ in thick Queensland kauri with clear epoxy finish. The sinks are 18-gauge No. 316 stainless steel, satin finish, with polypropylene traps and PVC wastes. The standard peninsular bench is 10 ft by 5 ft and has the following service outlets:

4 cold water over a trough; 2 cold water and 1 hot water over sink at end of bench
2 two-way gas; 1 single compressed air; 1 single vacuum;
4 10 A power (4 per circuit) and 1 15 A power (1 per circuit) on each side

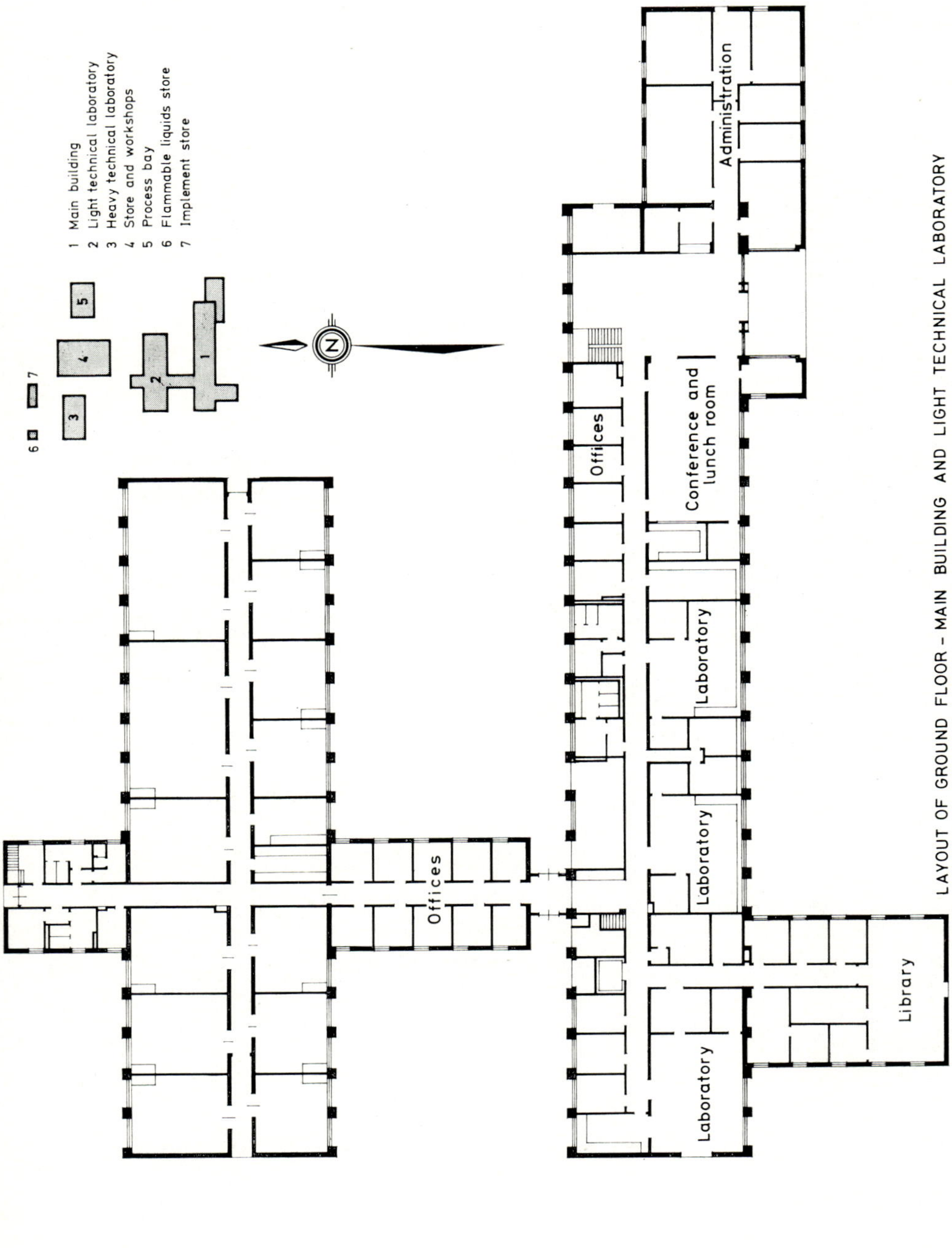

LAYOUT OF GROUND FLOOR – MAIN BUILDING AND LIGHT TECHNICAL LABORATORY

FIGURE 43 *Division of Chemical Engineering, Clayton, Vic.*

Air-conditioning

Thermostatically controlled fan coil units in the roof monitor, one per room; the humidity also is controlled; three changes of fresh air per hour. Fume cupboards are in rooms between laboratories; they are supplied with partially heated fresh air and have individual two-speed fans to exhaust 460 and 800 ft^3/min.

Services

Three gas-fired steam boilers, two 3400 lb per hour and one 1000 lb per hour at 250 lb/in^2; 1000 gal calorifier for domestic hot water and calorifiers to supply hot water for air-conditioning and space heating. Two Broomwade carbon ring air compressors 100 ft^3 at 100 lb/in^2, 25 hp motors, 40 ft^3 receiver. Two vacuum pumps, total of 250 ft^3 at 26 in of mercury, 20 hp and $7\frac{1}{2}$ hp motors, 35 cu ft receiver. Reticulated services are steam, hot and cold water, heating gas, compressed air, and vacuum. Copper pipe was used for the hot water, and galvanised wrought iron pipe for cold water, gas, air, and vacuum, with copper connections to the service outlets. One 1000 kVA 22 000/415/240 V transformer.

Fire protection

Sprinklers to main building and process bay; thermal alarms and smoke detectors to light technical and heavy technical laboratories and workshop; hydrants and fire extinguishers.

NATIONAL STANDARDS LABORATORY, BRADFIELD PARK, NSW

The National Standards Laboratory was established in 1940 on the campus of the University of Sydney. Plans are being prepared for the transfer of the Laboratory from the University to a 73-acre site at Bradfield Park in 1975. The Divisions of Physics and Applied Physics will be accommodated in 16 buildings which will have a total gross area of 435 000 $ft.^2$ The assignable space is as follows:

120 000 ft^2	—	General purpose laboratories
70 000 ft^2	—	Research offices, administration, library, lecture rooms, and cafeteria
35 000 ft^2	—	Ancillary service laboratories, workshops, and stores

The buildings are estimated to cost $14 million, and the building brief states that an allowance of $50 000 should be included in the estimates for the provision of appropriate works of art. The buildings will be of one and two storeys with brick load-bearing walls, and they are designed to accommodate a total staff of 450. The cafeteria will have a seating capacity of 270, and there will be provision for parking 250 cars. There will be three two-storey laboratory buildings each 320 by 100 ft with double-width layout—*i.e.* on each side of a 5 ft wide service corridor there will be a 24 ft wide laboratory, a 6 ft corridor, and a 12 ft laboratory office or service laboratory. The module has been increased from the generally accepted 10 ft to 12 ft to provide for wider benches with more space between. For protection from television and radio signals, reflective glass will be used in the windows and wire mesh in the external brick cavity.

Reticulated services will be hot and cold water, cooling water, town gas, and compressed air, with restricted reticulation of nitrogen, helium, and glycol-brine. Mechanical and electrical

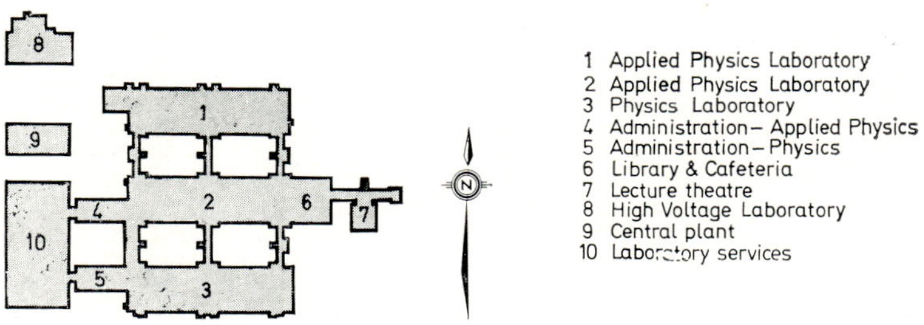

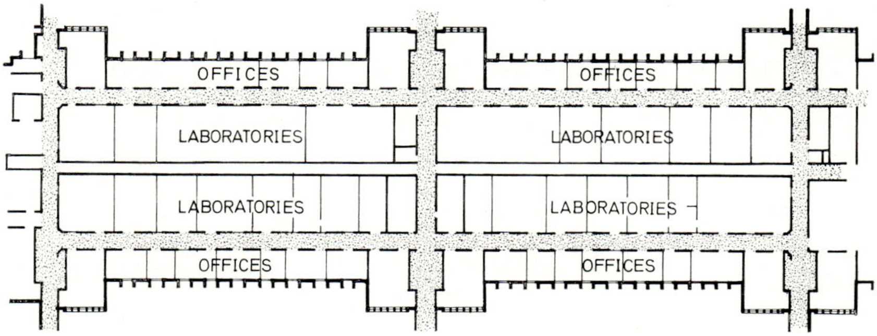

TYPICAL FLOOR PLAN

FIGURE 44 *National Standards Laboratory, Bradfield Park, NSW.*

equipment will be in a central plant building with a system of tunnels to all buildings. The brief provides for two hot-water boilers with a total capacity of 100% of the design load, and three equal-sized centrifugal water chillers with a total capacity of 120% of the design load.

The air-conditioning system has been designed to cope with altered layouts and conditions. On each floor of each laboratory building there will be eight plant rooms to serve an area of 320 by 98 ft—each plant room supplying air to a group of rooms, with separate ducts to each room. The inlet duct will be exposed and extend down the centre of each 12 ft module for the full depth of 24 ft. The return air grille will be above the door, and be fitted with a damper having a 1-hour fire rating; the corridor will be a return air plenum. The heat load is based on lighting of 50 lumens per square foot (2·8 W per square foot), one person per 120 ft^2, and equipment 1 kW per 144 ft^2. There will be three changes of fresh air per hour (or the sum of the exhaust air quantities plus one air change per hour for pressurisation, whichever is the greater), and 20 changes per hour so that, between the heights of 2 and 5 ft above the floor, the temperature and humidity will not vary more than ±1°F and ±5% RH. The operating temperature will be adjustable within the range of 68–74°F. Based on actual tests on a prototype laboratory, special control systems have been developed to meet the exacting requirements. A monitoring and control panel is to be installed to operate the 50 plant rooms.

Chapter 8

Overseas Laboratories

The basic problems associated with the design of research laboratories are the same in most countries, so it is interesting and often helpful to study as many designs as possible.

Since the 1930s, the United States of America and Canada have accepted flexibility as one of the most important design considerations; it is regarded as an essential in order to meet the demands of the constantly changing research requirements, and as an investment which will delay obsolescence. The layout of most laboratories is based on a module—generally 10 ft, but a few have 11 ft and some even 12 ft; most laboratories have a depth of about 24 ft. The provision of air-conditioning, more fume cupboards, and the reticulation of a greater number of services has meant that much more space was required for pipes and ducts. This has been provided by a series of vertical shafts or, in some laboratories, a service corridor which extends the full length of the building. Modular planning, movable walls, ample and accessible services, and unitised benches contribute most to flexibility, and nearly all modern laboratories in these two countries incorporate all these features.

When I visited laboratories on the Continent, my main impressions were the thoroughness with which they had been planned, the meticulous attention to detail, and the excellence of the workmanship. There was extensive use of modular planning and repetition of standard layouts, and plenty of services which were readily accessible; but flexibility in the sense of movable partitions and benches had not been considered necessary. Partition walls were brick, benches were not of unit construction, and bench tops were tiles set on concrete slabs.

This chapter is devoted to technical descriptions of five overseas laboratories. Obviously there are many outstanding laboratories, both large and small, of which I am not even aware; but, although not new, the five I have chosen incorporate features of particular interest, and all represent a practical and successful approach to laboratory planning.

ABBOTT LABORATORIES, NORTH CHICAGO, ILLINOIS, USA

Abbott Laboratories is a world-wide pharmaceutical manufacturing company with headquarters in North Chicago. The research facilities on this site were doubled when an eight-storey 225 000 ft^2 laboratory was built in 1960. Naess & Murphy, Architects and Engineers, Chicago, were responsible for the architectural and structural design, the mechanical and electrical services, and the supervision during construction. Details of this laboratory are:

> Reinforced concrete frame; 74 ft wide with a central pipe and duct service shaft and peripheral corridor. The column spacing is 20 ft to give a 10 ft module; the standard laboratories are 20 or 30 ft wide by 25 ft deep, and the floor-to-ceiling height is 10 ft.

Laboratories 20 ft wide accommodate a maximum of four persons; writing tables are provided for two senior research scientists, and smaller tables for assistants. Office space 25 by 10 ft for section leaders and secretaries is provided as shown in Fig. 46.

The internal walls of the corridor are demountable steel with clear glass above 3 ft 6 in; cross partitions are demountable steel frame with uncoloured asbestos cement or wired glass panels; walls between the laboratories and the service shaft are hollow concrete block. The floor is asphalt tile, and the underside of the concrete slab is exposed and painted.

FIGURE 45 *Abbott Laboratories, North Chicago, Ill., USA.*

The laboratory fittings are of standard design and a standard plan has been adopted; any variations do not alter the basic layout. The bench units are steel with $\frac{3}{4}$ in thick Colorlith bench tops. Reticulated services are hot and cold water, de-ionised water, steam, gas, compressed air, vacuum, and electric power. Water at a controlled pressure of 60 lb/in^2 and chilled brine are supplied from a central plant building. The de-ionised water is reticulated in polythene, and the outlets are silver-plated brass with Saunders diaphragm isolating valves. The waste pipes are Pyrex and the traps are lead. Details of the service outlets in a standard laboratory are:

Bench 15 ft long	—	Two groups each consisting of twin gas, air and vacuum, single cold water and cup sink, and four power.
Wash-up sink	—	Hot and cold water; de-ionised water.
Fume hood	—	Single gas, future bottled gas, two single air and vacuum, cold water and cup sink, steam and steam bath, three power.

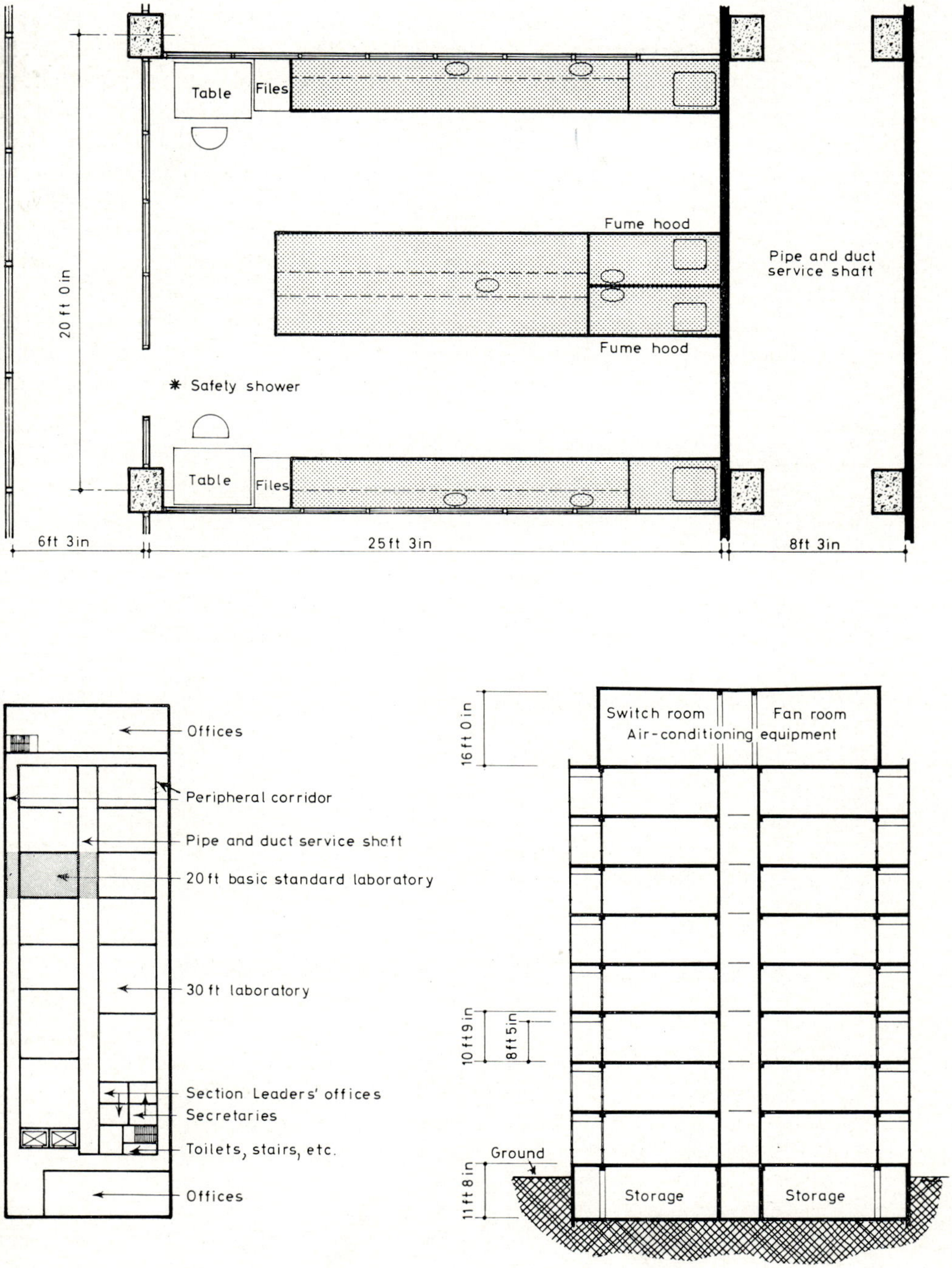

FIGURE 46 *Plan of basic standard laboratory layout, a typical floor plan, and a section through the 224 by 74 ft building—Abbott Laboratories, North Chicago, Ill., USA.*

The building is air-conditioned and the equipment (including supply and return fans, fume hood exhaust fans, and transformers) is in the penthouse. Inlet air enters the laboratories through registers high in the wall to the service shaft; it is exhausted through the fume hoods which operate continuously, and through registers low in the wall to the service shaft which also serves as a plenum for return air to the fans in the penthouse. Air also is supplied to the corridors through registers under the windows; this air flows across the corridors and into the laboratories. In order to reduce the loss of conditioned air through the fume hoods, make-up air is introduced at the top of the hood to supplement the hood exhaust volume; this is heated in winter, but not cooled in summer. No allowance was made for heat load from laboratory equipment; any devices small enough to be set on the bench top are placed in the fume hood if they pose a heat problem. Large floor-mounted items such as autoclaves are preferably grouped in separate rooms; if they must be used in the laboratory, each one has its own exhaust hood. In the laboratories there are 5–6 air changes per hour with 70% recirculation, and in the animal section there are 8–10 changes per hour with no recirculation.

Six years occupation proved the layout of the laboratories and services to be so satisfactory that the same type of layout has been adopted in another laboratory recently completed. The new building is twice the width and therefore has two pipe and duct service shafts, and an internal corridor. Partition walls are steel studs faced with both plaster and snap-in metal access panels to services within the stud space. Pyrex traps and piping are used throughout, including underground laboratory wastes. Fume hood ducts are asbestos cement or stainless steel. Instead of curtain-wall construction with continuous glazing, the new building has brick external walls with windows modest in size and number. Additional points of interest are:

There have been no problems with vibration since the large fans in the penthouse were balanced.
There has not been any corrosion in the steel bench units, and the Colorlith bench tops have been very satisfactory.
The capacity of one 50 A sub-board per 20 by 25 ft laboratory has been adequate.
Very few laboratory partitions or benches have been shifted; alterations have been greater in the office space.

BATTELLE INSTITUTE, FRANKFURT (MAIN), WEST GERMANY

The Battelle Institute is engaged in research and development in the biosciences, chemistry, metal science and technology, physics, silicates, mechanical and electrical engineering, and economics. This work is done on a contract basis for private industry, trade associations, and government departments; no subsidy is received. It follows, therefore, that the design of the buildings is studied closely to provide essential requirements and maximum flexibility at minimum cost and maintenance.

In his Activity Report 1966, the Director, Prof. Dr M. Barnick, states:

'Although success cannot be measured in figures, we should like to add the following data: In 1952/53 when our work was started, our staff totalled 19 members. In the meantime, the

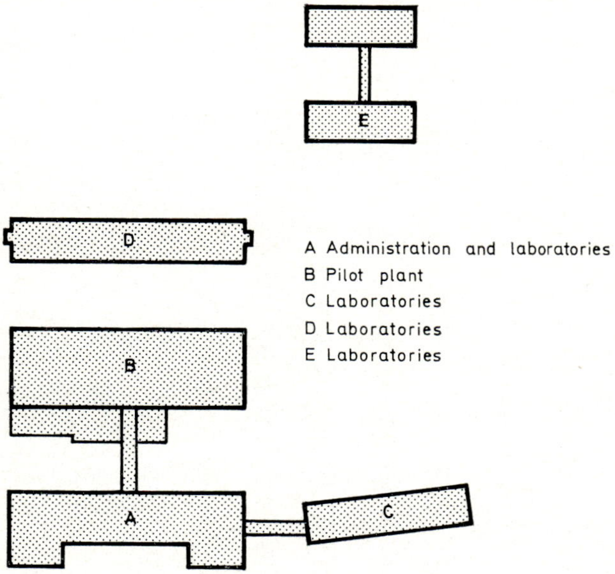

FIGURE 47 *Aerial view and key plan—Battelle Institute, Frankfurt (Main), West Germany.*

research expenditures increased from \$4400 in 1953 to \$6 000 000 in 1966, and the number of staff members rose to a total of 901. . . . Scientific research is becoming increasingly important, quite independent of the economic situation. We have to think today of what may happen tomorrow. In this respect contract research is an element of sound growth because it permits a company to conduct research even if the staff and equipment or funds required for internal research are not available. This applies in particular to medium and small-sized companies and to the investigation of complex problems for big firms. Problems are easier to solve if science and experience become partners. This is for us the most important element of contract research.'

Details of the main buildings erected on the 37-acre site are shown in Fig. 47. The module is 4 m, and the standard laboratory is 6 by 4 m with 5 by 4 m offices across the corridor. Improvements incorporated in the design of Laboratory D include:

Reduction in the depth of laboratories from 7 to 6 m because it was considered that the daylight was inadequate in the inner section of the deeper laboratories. The narrower laboratories also reduced the danger in case of fire.

A basement to provide space for the long-term storage of apparatus and equipment not in use.

Two windows per module because it was found that one window per module restricted the flexibility.

Pipe risers at alternate internal columns so that, in the event of an emergency, the supply of any service can be cut off from the corridor without entering the laboratory; furthermore, an exit door to the adjacent laboratory can be provided near the window. In previously built laboratories, the pipe risers were at alternate external columns.

The buildings are of reinforced concrete construction; external and internal walls are of 49 by 24 cm Ytong blocks which are light-weight concrete made from slag, lime, and sand to which

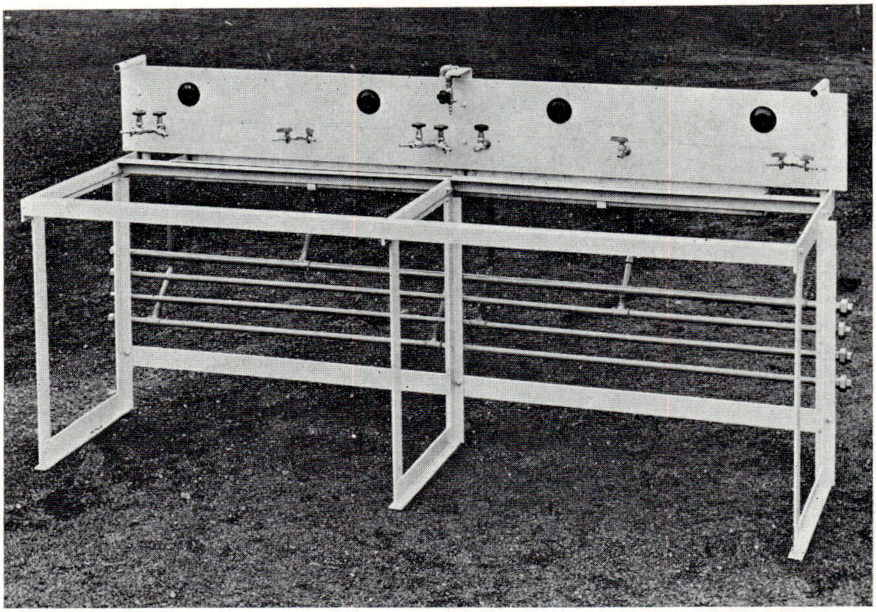

FIGURE 48 *Prefabricated bench frame—Battelle Institute, Frankfurt (Main), West Germany.*

aluminium powder is added as an expanding agent. These can be sawn, nailed, and drilled—a great advantage in view of the extensive modifications frequently necessary in these laboratories. External walls are rendered; internal walls are painted in the laboratories, offices, etc., and plastered and painted in the reception rooms. Doors open into the corridor; they are 1 m wide with lever latch. The floor-to-floor height is 3·6 m; the underside of the concrete slab is exposed and painted, and it has slotted steel strips (set in flush) for fixing equipment and light fittings. The floor is designed to take a heavy load (live load 1000 kg/m^2) in order to ensure that any required experimental equipment can be installed, *e.g.* lead shielding for studies in the field of nuclear chemistry. Openings in the floor slab at each internal column are provided for pipes and ducts; these holes are 30 by 30 cm on the ground floor, 30 by 60 cm on the first floor, and 60 by 60 cm on the second floor, and any not in immediate use are filled with weak-mix concrete.

Services reticulated are hot and cold water, gas, and compressed air. Electrical reticulation is by bus bars along the corridors, connected centrally in the length of the building so that voltage

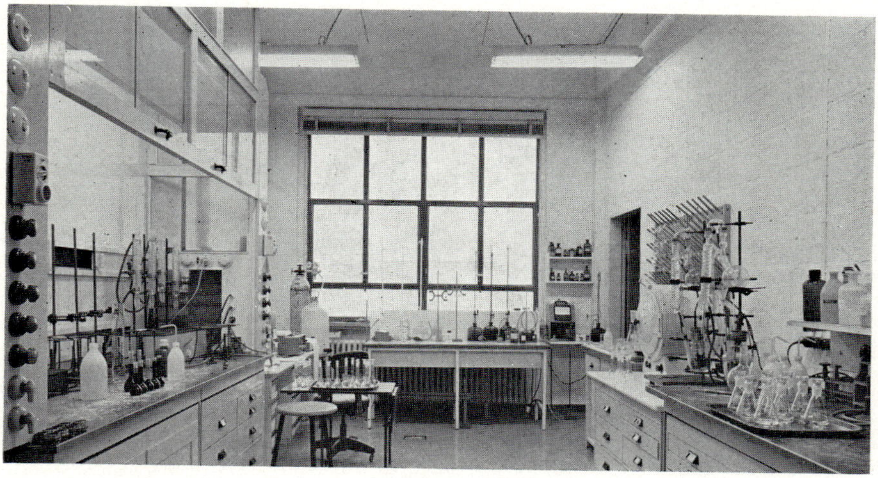

FIGURE 49 *Fume cupboard—Battelle Institute, Frankfurt (Main), West Germany.*

drop is a minimum. The plug-in feeder boxes are provided with fuses for 25, 60, and 100 A, and they can be controlled at intervals of 25 cm.

Standard bench units have steel frames 2·5 m long; each section is prefabricated complete with service pipes and electrical wiring, a back skirting on which the outlets are mounted, and a trough extending almost the full length—as shown in Fig. 48. Each unit is fitted with five cold water outlets, one hot water, four gas, one compressed air, and four power outlets. Three types of removable timber drawer and/or cupboard units are available for use underneath. The bench top is 2 cm thick compressed asbestos cement with a waxed finish.

The fume cupboards have vertically sliding sashes, the lower sections of which are fitted with horizontally sliding panels of laminated glass; the ducts and exhaust fans are PVC. Normal service outlets are within the cupboard, with controls outside. Power outlets with hinged covers are inside the cupboard, but the switches are outside. The box below the three switches (*see* Fig. 49) is a 'current tap' for controlling the electric heating or for adjusting the speed of the agitators.

HOECHST RESEARCH CENTRE, FRANKFURT (M)-HOECHST, WEST GERMANY

The Farbwerke Hoechst AG is a firm with a world-wide reputation for chemical products; headquarters are at Frankfurt (M)-Hoechst. In the course of a century of development, the research laboratories were widely dispersed amongst the Hoechst factories; this was uneconomic and had the serious disadvantage that research workers were in isolated groups. On this account, in 1959 it was decided to establish a research centre on a 74-acre site on the opposite side of the River Main and connected to the existing site by a company-owned bridge. The Company's Division of Buildings and Laboratory Construction was responsible for preliminary planning, design of laboratory fittings and services, and supervision. A number of leading architects were invited to prepare designs and a master plan, and the scheme submitted by Professor Henn of Braunschweig was accepted.

On the occasion of the official opening of the first laboratory in September 1960, 'Hoechst News' interviewed the people responsible for the construction of the building—Dr Schultheis, member of the Executive Board, Mr Friedrich, and Mr Grueneis. Extracts from the interview are:

> *Dr Schultheis*—The selection of this terrain was governed primarily by its special situation and rural seclusion which coincided with another aim, namely to bring our scientists closer together so as to intensify their intellectual contacts. We want to give them an opportunity of being able to exchange their ideas and experience also outside the regular meetings and discussions and away from the noise and bustle of production.
>
> *Hoechst News*—Isn't there a chance that separation from production, particularly in an industrial undertaking, might have adverse results? Does not the research worker also need direct contact with production?
>
> *Dr Schultheis*—Certainly he needs this contact, and we have, of course, not interfered with it. But it was important for Hoechst to draw a sharp line between the tasks of research and those of production. In any case, every research worker must try out his laboratory results on the smallest technical scale. We have now provided him with this possibility by equipping each laboratory building with a pilot plant. In many cases, these pilot experiments will be adequate to establish the conditions, in respect of both methods and materials, necessary for large-scale manufacture.
>
> *Hoechst News*—The main laboratory, which has now been completed, gives the impression of being a well-planned building. What are the fundamental ideas on which the design has been based?
>
> *Mr Friedrich*—Because of the location of the site, we have been in the fortunate position of being able to plan generously and of being able to realise the wishes of the Company. Nevertheless, we employed the utmost economy and calculated our space requirements exactly. The basic unit in our planning was the two-man laboratory. We constructed a prototype of this and studied it closely.
>
> *Hoechst News*—Why did you choose the two-man laboratory?
>
> *Dr Schultheis*—We had to decide whether to build a one-man laboratory or whether to provide space for larger research teams. Our long experience led us to the view that a laboratory designed for two chemists is, for many reasons, the most practicable unit. The chemist can work more quietly than in a large laboratory and is able to evolve his own approach. The one-man laboratory, on the other hand, has the disadvantage that it affords no facilities to its inhabitant for exchanging views with his colleague so that he may tend to become something of a hermit. In addition, we intended by the two-man principle to ensure that the laboratory is continuously manned, particularly during critical operation, so that a high degree of safety is achieved.

Hoechst News—How many individual laboratories can be housed in one unit?
Mr Friedrich—Each laboratory unit has 36 identical laboratories, together with the requisite writing and ancillary rooms. The space arrangements permit the setting up of teams in whatever size required, working under the direction of specialists.

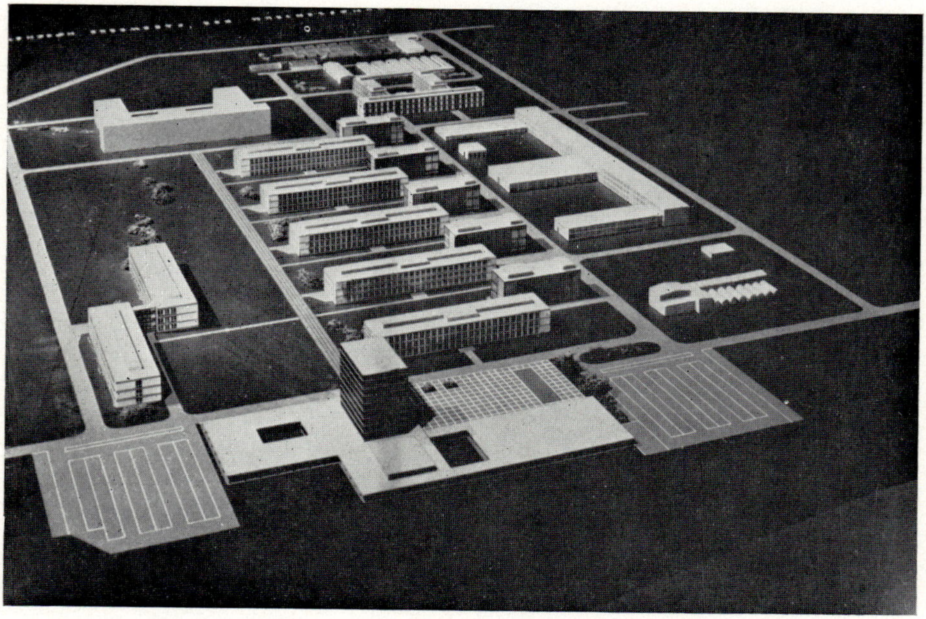

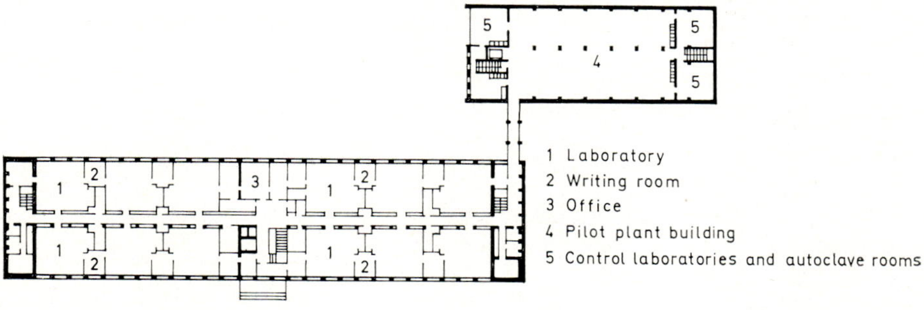

FIGURE 50 *Model of the site development, and a typical floor plan of a* 101 *by* 22·5 *m laboratory and* 48·75 *by* 18·75 *m pilot plant building—Hoechst Research Centre, Frankfurt* (*M*)*-Hoechst, West Germany.*

Hoechst News—Why did you depart from the modern tendency to build skyscrapers? After all, land is precious even if in this case there was no shortage of it.
Mr Friedrich—Unless limited space makes them absolutely necessary, skyscrapers appear to us for a number of reasons unsuitable for chemical laboratories. The most important of these is that regulations and bye-laws governing the construction of skyscrapers are difficult to meet in these cases. In addition, in a skyscraper more cubic space is taken up by the ventilating system, service lines, stair wells and lift shafts than in a low building. This is

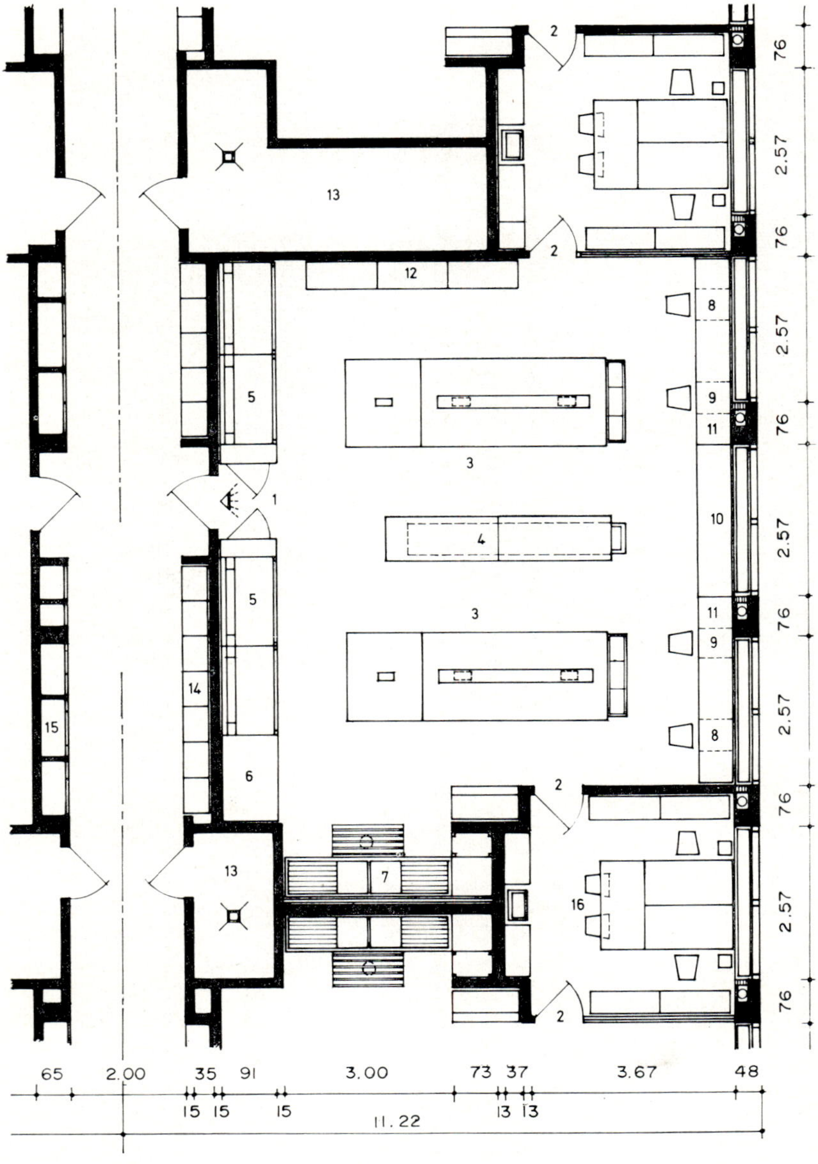

1 *Entrance with safety sprinkler*

2 *Emergency exits and entrance to writing room*

3 *Island bench*

4 *Temporary bench with services in floor compartment*

5 *Two fume cupboards with adjoining cupboard for steel cylinders*

6 *Drying ovens and ventilated chemical cupboard*

7 *Two-bowl sink unit with ventilation and floor waste*

8 *Writing space for laboratory assistant*

9 *Window bench*

10 *Balance bench*

11 *Reagent shelves*

12 *Cupboards*

13 *Special room*

14 *Wall cupboard in corridor for chemicals and glassware*

15 *Ventilation shafts*

16 *Writing room for two chemists*

FIGURE 51 *Layout of a basic laboratory—Hoechst Research Centre, Frankfurt (M)-Hoechst, West Germany.*

because the ventilation and air conditioning shafts from each section must be separated by fireproof walls and conducted separately over the roof into the open. In low buildings, these ducts can be easily fitted between the supports without taking up valuable laboratory space.
Hoechst News—Has each laboratory building its own pilot plant?
Mr Grueneis—That is so. All laboratories and pilot plants are connected by enclosed passages. Each laboratory building and its pilot plant together form a coherent structural unit. There

FIGURE 52 *A typical laboratory—Hoechst Research Centre, Frankfurt (M)-Hoechst, West Germany.*

are five of these on the new site. Between the individual buildings there is the much-needed space for open air installations which often is missing in the case of tall buildings. The various laboratories receive their supplies by way of underground passages. Electric trucks take them to each floor with the aid of heavy goods lifts.

Details of the buildings are as follows:

There are five chemistry laboratories each 101 by 22·5 m; these are for fundamental, synthetics, dyestuff and fibre, and drug research. Each laboratory has a pilot plant building 48·75 by 18·75 m. All these buildings have three storeys and basement; they are interconnected by enclosed corridors which have two storeys and basement. In addition there are six laboratories of similar size specially designed for physics, analytical chemistry, pharmacology, and engineering research. Details of the master plan are shown in Fig. 50.
The buildings have a prefabricated reinforced concrete frame and brick external walls; the internal walls are brick, plastered, with fairly extensive areas tiled. The columns are at 3·33 m centres, the floor-to-ceiling height is 3·2 m, and the floor-to-floor height is 4·2 m. There is one window 2·57 m wide per module of 3·33 m. There are holes in the cross-beams for pipes and ducts. The suspended floors of the pilot plant buildings are precast concrete slabs set in Asplit cement and supported on steel joists which are bolted to the columns;

therefore the whole or portion of the floor can be removed, or one slab can be replaced by another with a hole or holes in the required positions.

Exhaust fans and lift machinery are housed in the roof space of the laboratories. In the basements of the pilot plant buildings are pump and transformer rooms, a brine-producing unit, acid stores, ice stores, and a large space for equipment not in use. Also air-conditioned solvent stores are in the basements between pilot plants and laboratories.

Bench units are timber, of standard design, and fixed to timber frames; they are set on a concrete base with cove to floor. The bench tops are plywood, sealed with Asplit CN and covered with acid-proof tiles set in Asplit cement. Each 4·6 by 1·56 m island bench has 14 single-phase and four three-phase outlets. Services reticulated from the parent site include high and low pressure steam, compressed air, burner gas, nitrogen, and drinking water. Service mains are installed under each floor and pass through holes in the concrete beams as shown in Fig. 26; waste pipes are polythene. Between the island benches there is a temporary bench with service outlets (including power) in a floor compartment. Provision also is made in both floor and ceiling for fixing 2·5 cm diameter stainless steel rods to form a framework for mounting apparatus.

The electricity supply of 60 kVA to each three-module laboratory has proved adequate.

The ventilation plants are equipped with air washers so that in summer the room temperature can be noticeably reduced without having to use refrigeration, while in winter excessively dry air is avoided; there are 20 air changes per hour and no recirculation. The air enters the laboratory through nozzles which are installed in the sound-absorbing ceiling—50 per square metre over the entire area of the room. This ensures uniform distribution; furthermore, the velocity of the air passing through the nozzles is increased from 30 to 60 m per minute so that it reaches right to the floor. This draught-free ventilation and the resultant even temperature distribution has proved of great benefit to temperature-sensitive precision instruments and analytical equipment such as gas chromatography apparatus.

Air is exhausted through the fume cupboards into polythene ducts and vertical shafts which are polythene encased in brick. In each three-module laboratory there are four fume cupboards; each has a volume of 2 m^3 and, at 500 air changes per hour, this totals 4000 m^3 per hour; to this must be added 1500 m^3 per hour for spot ventilation over benches, wash-up sinks, and drying cupboards. The total quantity of air exhausted therefore is 5500 m^3 per hour.

The fume cupboards are 1·5 m wide by 1 m deep. The bench top is tiled with a raised edge and there is a ventilated cupboard beneath. The front sash slides vertically and has horizontally sliding panels in the lower section. Services include three single-phase and one three-phase power outlets, and remote-controlled outlets for water, burner gas, compressed air, vacuum, nitrogen, and distilled water. The sink is 26 by 11 cm. Fumes are exhausted through a slot formed by the back baffle which runs the full length of the cupboard. This slot is 30 cm above the bench and 11 cm deep. The back baffle has an opening in the top part of the cupboard so that either top or bottom exhaust, or both, can be obtained. The four fume cupboards are exhausted by a single fan with a polythene impeller and a capacity of 14 000 m^3 per hour. The fumes are discharged vertically through the roof.

The approximate cost, including design fees, was $194 per square metre.

In each three-module 10 by 9 m laboratory and 4 by 4 m office there are nine people—*i.e.* two research officers each with one technical officer, two assistants, and a cleaner half time.

Details of the safety precautions are:

Safety equipment located at the entrance doorway of each laboratory consists of hand fire extinguishers, fire extinguishing blanket, emergency shower, and eye bath—*see* Fig. 53. Isolating valves (accessible from the corridor) are fitted to flammable gas services; electric

FIGURE 53 *Entrance doorway to a laboratory showing hand fire extinguisher, fire blanket, emergency shower, and eye bath; portion of a fume cupboard in the service shaft can be seen at the left—Hoechst Research Centre, Frankfurt (M)-Hoechst, West Germany.*

power also can be switched off from the corridor, but, for safety reasons, not the electric light.

Isolating valves and switches are fitted to all services so that they can be shut off by floors.

In addition to the main door, each laboratory has at least one emergency exit—a fire-proof connecting door to the writing rooms of the adjoining laboratory. This is accessible from any position; for this reason, only island benches were used.

On each floor of every chemistry laboratory there is a bath of water kept constantly at body temperature for cases of emergency with acid burns, etc.

From a central control point, inlet and exhaust air registers can be shut off separately.

Fire hydrants, with water at controlled pressure, are installed throughout the site.

The Research Centre has been planned on the basis of a number of separate buildings rather than one large building, thus reducing the extent of damage in the event of a fire.

HOMER RESEARCH LABORATORIES, BETHLEHEM, PENNSYLVANIA, USA

The Homer Research Laboratories are built on a 70-acre site overlooking the Corporation's steel plant and offices. Surrounding this site, Bethlehem Steel has acquired about 1000 acres of timbered land on which a reafforestation programme is in progress.

In 1955 a member of the Research Management Section was appointed to assemble details of laboratory requirements for the approval of the Director of Research, and preliminary planning commenced. Architects for the project were Voorhees, Walker, Smith, Smith & Haines; they also were responsible for the design of the structural, mechanical and electrical services, and site planning. Construction was supervised by the Corporation with the assistance of two field engineers provided by the architects.

These laboratories concentrate in one location most of the research activities which formerly were scattered around the country. Several buildings were occupied in 1961, and since then expansion of facilities has been continuous. Most buildings can be extended—including an additional laboratory wing and two office wings to the main building—and there is space for several more buildings. If still more space is needed, there is the 1000 acres of undeveloped land!

The main building is three-storey and has five wings; it is occupied as shown in Fig. 54. The laboratory wing is 315 by 66 ft, and a total of 155 laboratory modules (each 24 by 10 ft) are located on both sides of a 7 ft wide corridor. Offices 15 ft wide are provided on both sides of a 7 ft wide corridor in an adjoining wing. Further details are:

Steel frame; reinforced concrete floors designed for live load of 150 lb/ft^2 in laboratories and 80 lb/ft^2 in offices. External walls are limestone with concrete-block backing. The internal column spacing is 20 ft to give a 10 ft module. The floor-to-ceiling height is 10 ft 4 in, with 2 ft 8 in between it and the floor above. In each 10 ft module there are two windows each 3 ft wide in the laboratory wing, and one window 4 ft wide in the office wing.

The internal partitions generally are demountable insulated steel panels, flush with posts at 4 ft centres; brackets for shelving and wall cabinets can be fitted to the posts. The floor is vinyl asbestos, and the ceilings are perforated steel panels in the laboratories and mineral acoustic tiles in the offices.

Scientists have individual offices in a separate wing, or an office area at the window side of the laboratory—depending on the type of work being done, the amount of time they spend in the laboratory, and the laboratory environment (*i.e.* whether or not it is clean, quiet, and relatively uncluttered). Writing space for assistants is provided either at the bench or at tables in the laboratory.

The laboratory benches are steel; the bench tops are soapstone, with timber, stainless steel, compressed asbestos cement, or linoleum used occasionally as required. Reticulated services available at each module are hot and cold water, distilled water, low pressure steam, natural gas, compressed air, vacuum, special gases, and 120, 208, and 460 V power. A 100 A 120/208 three-phase four-wire panel serves two laboratory modules. There is space in the service closets for installation and connection of compressed gas cylinders, as shown in Fig. 25; but cylinders holding toxic or flammable gases are clamped to the laboratory benches in the conventional manner. The distilled water pipes are PVC and the taps are tin-lined brass,

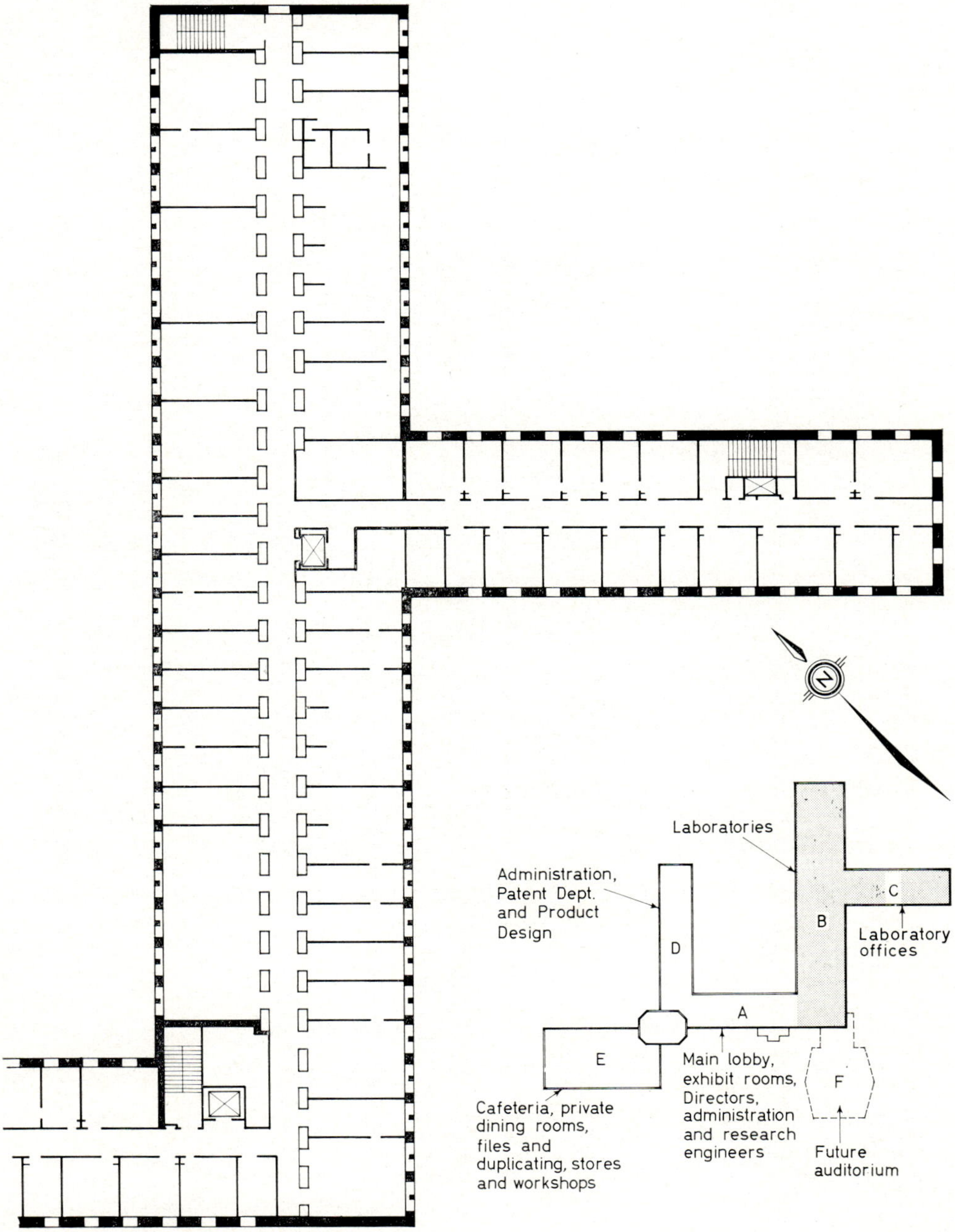

FIGURE 54 *Typical floor plan of Buildings B and C—Homer Research Laboratories, Bethlehem, Penn. USA.*

chrome plated. Air-conditioning equipment, exhaust fans for fume hoods, electricity bus ducts, and distilled water equipment are in the penthouse; sub-stations and vacuum pumps are on the ground floor, and compressors and boilers, etc., are in a central plant building; water pressure is controlled.

The rated capacity of the sub-station is 10 000 kVA. The site distribution system (a dual service at 12 kV) carries power to load centres in the buildings where it is distributed to panels at 480/208/120 or 277 V.

FIGURE 55 *A typical laboratory showing emergency exit door—Homer Research Laboratories, Bethlehem, Penn., USA.*

Laboratory effluent is drained into a 14 000 gal equalisation tank which also acts as a settling and, to some extent, self-neutralising tank; it is designed for a 100 gal/min flow and 60 000 gal/day based on a 10-hour flow and a detention time of $2\frac{1}{2}$ hours.

Where flammable material is being used, fume hood ducts are stainless steel; where perchloric acid is used, the ducts, fans, water spray piping, and nozzles are PVC. The fan discharge is vertical—almost flush with the penthouse roof. In general, fans and housings are steel with an acid-resistant coating.

Air-conditioning in the laboratories is of the dual-duct type, supplying conditioned air to under-the-window mixing boxes; air is exhausted either through the fume hoods or through ducts above the corridor ceilings; there is no recirculation. Offices are heated and cooled using under-the-window induction type units; there are 15 air changes per hour.

NATIONAL BUREAU OF STANDARDS, GAITHERSBURG, MARYLAND, USA

The National Bureau of Standards, US Department of Commerce, was established in 1901 on a 7-acre site in Washington, D.C.; adjoining sites subsequently were purchased, bringing the total area to 67 acres. By 1956 the staff had increased from 13 to about 2800 and a new site of approximately 550 acres was acquired at Gaithersburg, 20 miles from Washington.

In order to provide a reasonable involvement of the employees in the planning of the new facilities, a seven-man Laboratory Planning Committee was formed; its purpose was to recommend some of the more important criteria upon which the final plans were to be based. The Committee remained in existence throughout the design and construction period to consider special problems on which management needed advice. Virtually all of their recommendations were incorporated, although one of the important exceptions was the module. The Committee recommended 12 ft, and the architect 10 ft; but 11 ft was finally adopted. In their report to the Director, the Committee stated:

> 'Our views are based on our own collective experience in this and other laboratories, discussion with our colleagues at NBS, and special trips to eleven new laboratories. . . . Members of the Committee inspected these labs in the company of friends who worked in them, or in some cases with members of the managerial or planning staff. We were invariably treated with great courtesy by both management and staff, and were always given full information, often including a complete and candid discussion of their mistakes.'

This is an excellent report and, as many of the views expressed in it are very thought-provoking, I feel it well worthwhile to quote liberally from the Committee's recommendations:

Core Facilities
The library, a central auditorium and several small lecture rooms, the central stock rooms, the cafeteria, and the technical shops should be located in a central building easily accessible by foot to the technical areas. . . .

Library
The main library should be on the ground floor (not the roof) and should contain a large number of carrells which can be reserved, with smoking permitted. . . . Every effort should be made to make the library as easy to reach and to use as possible. . . . In addition to the carrells, there should be a room containing duplicating equipment adequate to the task of copying large numbers of single articles in journals. . . . Consideration should be given to providing an outside entrance to the library, so that it could more easily be made available at night without opening up the entire central building. . . .

Dining Rooms
The cafeteria should have a number of small dining rooms, seating up to about twenty people, to be reserved for luncheons with technical guests. . . .

Expanded Shops
We should provide expanded shops for glass work, instrument loan, electronics construction and repair and maintenance, and special fabrication services such as electronic tube work. An

extrapolation of past experience shows that scientists will continue to make ever-increasing use of more complex apparatus, and will be able to do less of this fabrication themselves. . . .

Administration

If all the administration is centralised in a multi-storey building, efficient planning suggests that the first floor should house the technical facilities (auditorium, library, storerooms, etc.). The second floor should contain the supply division and the 'line' administrative offices (dealing directly with division and section cheifs); and higher floors the administrative divisions in decreasing order of contact with technical staff and the general public. . . .

Storage Space

Scientists in almost all of the new labs we visited complained of lack of storage space. Attics should be subdivided into well illuminated rooms which can be locked. In addition, the modules should be spacious with partitions strong enough to support heavily loaded cabinets.

Conference Rooms

Most of the new labs we visited have too few conference rooms, and NBS is desperately short of them now. One new lab of 105 professionals has three conference rooms and an auditorium and finds this 'grossly inadequate'. Each divisional unit should have a staff lecture room, and each division should have at least two conference rooms (including one attached to the Division Office) as a bare minimum.

Shops

'Laboratory' and 'area' instrument, machine, glass blowing, and electronic shops, photographic laboratories and similar facilities must be retained at the section or inter-sectional level, although in some cases central divisional facilities may be permissible. The basic laboratory units, if properly designed, will also accommodate these shops, since they normally contain relatively light machine tools. At least one glass blower, mainly for repair and installation work, should be located in each of the main building groups housing physics and chemistry activities.

Vibration

With knowledge of the NBS preliminary plans to build basementless laboratories we specifically looked into vibration problems in other laboratories. Bell and Westinghouse have heavy rotating machinery in basements and scientists there suggested it be housed in separate buildings. Scientists at these and at all of the other labs were horrified at the thought of deliberately placing machinery in the attic. This problem is apparently particularly severe in steel frame buildings. . . . Throughout the new laboratory the philosophy on nuisances, mechanical or otherwise, should be to isolate the source rather than try to shield everything else. It is cheaper to isolate the source of vibration than to shock mount all the apparatus.

External Doorways

Every building freight elevator must be accessible to the interbuilding tunnels and to a street level loading platform so that delicate and heavy equipment can be moved without encountering steps. . . .

Building Services

Each building should contain one or more storage and dispensing rooms for dry ice, liquid air, organic solvents, bottled gases, and other items needed constantly, on short notice, and at night.

Module Size

We consider our most difficult problem encountered in this study was the selection of the best module size and shape, and we gave a large fraction of our attention to this question. We start from a premise upon which we all agree: a large number, perhaps a majority of the scientists will want separate, quiet offices and should have them. Most scientists cannot think in a laboratory full of machinery and technicians any more than an executive can think in a room full of typists. The number of such offices which will be required will vary greatly from section to section and from time to time. Further, it is a matter of past experience in all laboratories that a new laboratory starts with the optimum distribution of lab and office space, expands at the expense of the office space until the latter is largely consumed, and then a new laboratory is built. It is virtually inevitable that some day scientific office space will be converted into laboratory space. Hence we conclude that we must have scientific offices, of variable number and location, and potentially convertible into laboratories.

This leads us to reject the off centre corridor design with a small office module on one side, in favor of a symmetrical wing with potential laboratories on both sides. This design permits two methods of providing desk space, providing the basic module is large enough. Either the space may be partitioned into small individual offices, or a corner of the laboratory can have a half-partition. In the latter case, a 10 ft by 26 ft module will be too small. Therefore we favor a 12 ft by 24 ft or 12 ft by 26 ft module. It could house one physicist, his assistant and a rather large apparatus, or perhaps two chemists. . . . Ceiling height should everywhere be held to a 10 ft minimum, and in physics space should be 12 ft.

Service Shafts

The Committee accepts . . . vertical service shafts between the corridor and the lab modules. However, several of us were greatly impressed by the 'inside-out' arrangement at the National Carbon laboratory. The services there are distributed in a horizontal corridor adjacent to which two rows of labs are arranged, back to back. The personnel corridor then surrounds the two blocks of labs. Completely windowless laboratories result. The convenient and inexpensive distribution of utilities in this design are far superior to the vertical shaft design. In addition, bulky and noisy pieces of equipment, such as forepumps and voltage regulators, can be placed in the corridor. . . .

The service shafts themselves should be more spacious at NBS than at comparable other laboratories because the tremendous diversity of NBS activities requires a large amount of special purpose wiring and plumbing. The shafts should also have enough space in them so that large (*e.g.* 8 in) vacuum lines might be run from some laboratories to the basement, where the pump would be installed. As much of the service shafts should be exposable as possible, by removing panels from the laboratory walls, so that only the ceiling to upper-floor thickness is obscured.

Laboratory Doors

Standard 3 ft doors should be swung from a fly-door about 18 in wide which can be opened to provide $4\frac{1}{2}$ ft of clearance when needed. . . .

Ventilation

The buildings should be centrally air conditioned and every effort should be made to see to it that this expense is not the one most easily cut at the last minute. Air conditioning in laboratories containing equipment dissipating 5–10 kilowatts of heat is a necessity and not a luxury. Humidity and temperature control often are necessities. . . .

Strong drafts from the ducts should also be avoided. In the chemistry building each room should have an independent exhaust stack, connected to the hood. In physics space a group

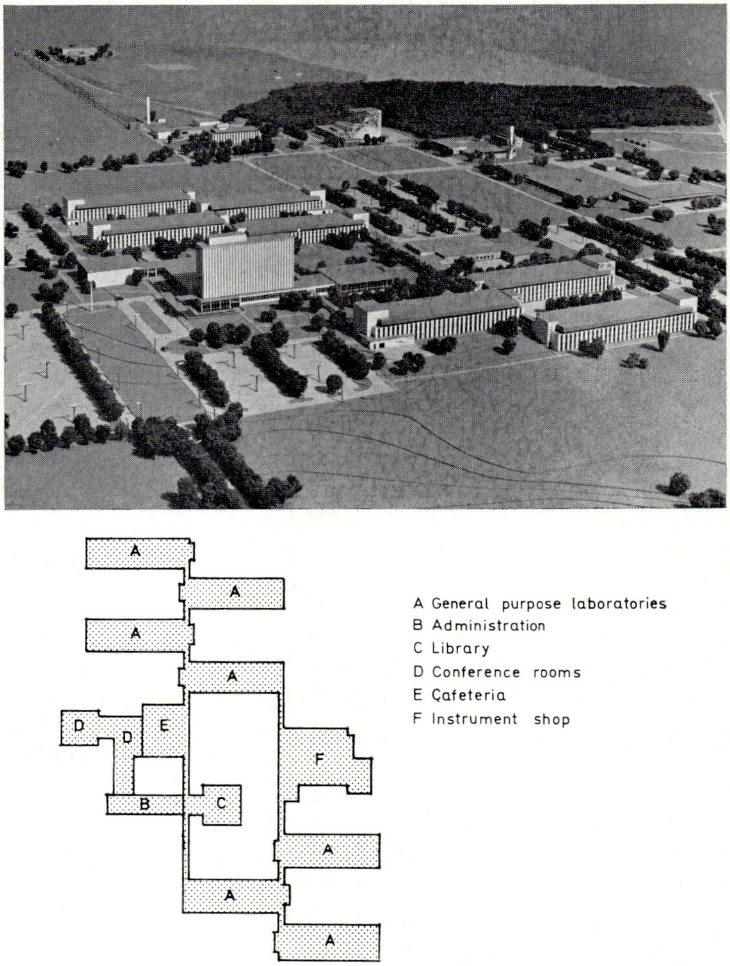

FIGURE 56 *Model of the site development, and key plan—National Bureau of Standards, Gaithersburg, Md., USA.*

of about four modules might share a single exhaust vent. Laboratory air cannot be recirculated, at least in chemistry space. Special attention should be given to the manner in which the exhaust air is discharged to the outside. At National Carbon they went to great trouble and expense to provide independent air conditioning systems for two spectroscopy laboratories requiring very clean air. They discovered that the exhaust of the other chemistry labs came into these special intakes and created an intolerable situation. . . .

Plumbing
Every laboratory should have installed burner gas, compressed (oil-free, dried, and filtered) air, hot and cold filtered water, and drains. Every service shaft should contain three additional gas pipes; nitrogen, oxygen, and a third which would be filled with hydrogen in some divisions, perhaps argon or house vacuum in others. These facilities should then be piped into all the labs desiring them. This will probably include at least half of the laboratories, and these gases should be available to the others when needed. The gas distribution system would greatly increase convenience in research, add space to the laboratories, remove the safety hazard of large high-pressure tanks standing in the lab, and cut the inventory and demurrage costs.'

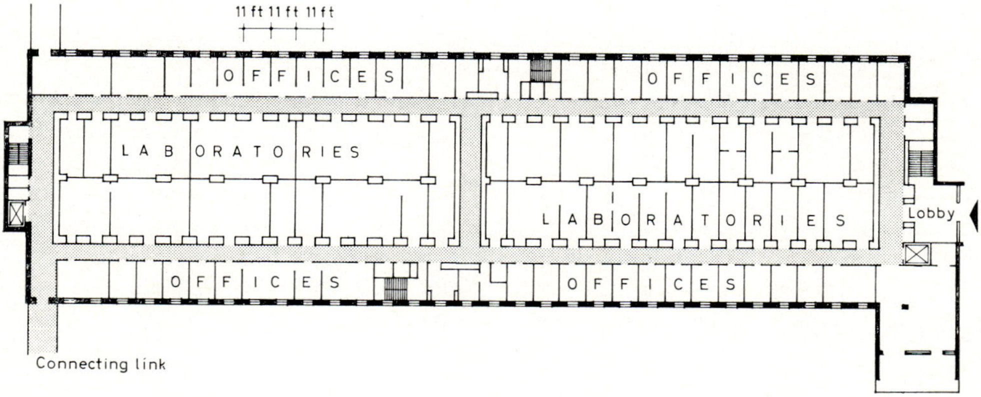

FIGURE 57 *Typical floor plan of a* 105 *ft by* 385 *ft general purpose laboratory—National Bureau of Standards, Gaithersburg, Md., USA.*

The New York firm of Smith, Haines, Lundberg and Waehler—then known as Voorhees, Walker, Smith and Smith—was selected to design the architectural, mechanical, and electrical work. General Services Administration was responsible for placing contracts and for supervision during construction.

The architects studied various schemes and finally concluded that a single building to accommodate the main laboratories, administration, and workshops would be exceedingly big and cumbersome, and would not lend itself well to future expansion. It was decided, therefore, to have a central administration building surrounded by a number of general purpose laboratories. In addition, there are separate buildings for Radiation Physics Laboratory, Nuclear Reactor, Engineering Mechanics Laboratory, Hazardous Chemical Laboratory, Industrial Engineering Building, Acoustics Laboratory, Concreting Materials Building, Fluid Mechanics Laboratory, Non-Magnetic Laboratory, and Boiler Plant. The gross area of all buildings totals 2·3 million ft^2; the assignable area is 1·2 million ft^2—*i.e.* the average utilisation factor is 52%.

There are seven double-width general purpose laboratories for physics, chemistry, and engineering research; each of these buildings is 105 by 385 ft, with the long elevations facing north and south; they have three storeys (three also have basements) and are connected with three-storey covered ways. Site planning provides for seven more similar buildings. The buildings as handed over by the contractor were based on working requirements laid down almost five years earlier.

Before they were occupied, approximately half the laboratories required some modifications to space or service outlets. The relative ease with which these changes were made demonstrated the value of flexibility. Details of the buildings are:

> Reinforced concrete frame; reinforced concrete floors designed for live load of 150 lb/ft^2; brick external walls. The column spacing is 22 ft to give an 11 ft module; widths of offices 16 ft, corridor 7 ft, laboratories 24 ft, with vertical service shafts where shown in Fig. 57. The floor-to-ceiling height is approximately 11 ft. There is one 5 ft wide window per module. The basic laboratory is 11 ft by 24 ft and generally accommodates one scientist and one assistant. Smaller rooms 11 by 16 ft on the opposite side of the corridor are intended primarily as laboratory offices although they can, with certain limitations, be used as laboratories. They are occupied by one section leader or two scientists.
>
> There is a significant amount of highly specialised space—for example:
>
> > The tape calibrator room in the basement of the Metrology Laboratory is 320 ft long, 11 ft wide, and 8 ft high. It was built underground in order to achieve a more constant and uniform distribution of temperature. Further temperature stability is provided by a double wall construction with the cavities filled with insulating materials, and refrigerator-type doors at each end of the tunnel.
> >
> > In some areas of the building research laboratory, floors have been omitted to provide space three storeys in height.
>
> The internal walls generally are demountable steel. The floor is vinyl asbestos tile, and the underside of the concrete slab is exposed and painted in the laboratory areas; acoustical tile is applied to the concrete slabs in perimeter offices.
>
> The laboratory fittings are of standard design; the bench units are steel, and the bench tops are birch or maple for dry work, and soapstone for wet chemistry. Reticulated services available at each module are hot and cold water, steam, distilled water, chilled water for heat exchanging, burner gas, compressed air, and special gases where required. Steam, chilled water, and compressed air are supplied from a central plant building. The waste pipes are tempered glass.
>
> Each building is air-conditioned and the equipment (including supply and return fans, water stills, and transformers) is in the penthouse. Air at about 50°F is supplied to the outer (office) rooms in four zones and passes through a low-pressure induction unit below each window, and the final temperature can be adjusted to meet individual needs. Air is returned in a duct in the corridor and is recirculated; some can infiltrate laboratories for fume hood make-up. The system for internal laboratories is different in that each 11 ft module has its own supply air duct from the penthouse; the hot water reheat units also are in the penthouse, with thermostatic control in each laboratory. Air is returned either by duct (in which case it is recirculated) or through the fume hoods when it is exhausted direct by duct through the roof. Air independent of the air-conditioning system is not supplied to fume hoods. This system has been designed to give flexibility and reduce the chance of cross-contamination in critical areas. In some rooms, perforated ceilings or run-out ducts with diffusers were used to improve air distribution and reduce velocity. In addition to the normal temperature and humidity controls and filtration, the environment in each room can be further modified by the installation of booster efficiency filters, additional chillers, humidifiers, coils, etc.

Electrical

Each building is served by three 13 200/480/277 V transformers of 500, 1000, and 1500 kVA capacity. These transformers feed the 277 V lighting circuits and the 480 V power systems, and they also feed the two 480/120/208 V transformers located at the mid-point of the penthouse. The latter two transformers supply all 120/208 V service, and feed individual cables to alternate vertical shafts serving laboratory areas on the floors below; thus the transformers which provide power for laboratory use are isolated from those for lighting and mechanical services. Voltage variations in excess of 2% are overcome by individual regulators in the laboratories. A 100 A panel serves each group of four modules, two internal laboratory, and two office or limited laboratory areas. The light level was set at 65 lumens per square foot. Some lighting groups recommend a higher level of illumination, but the Bureau's own lighting experts concluded that the difference between 65 and 100 lumens per square foot was not really worth the additional cost in electric power or in the additional air-conditioning that would be required. Fluorescent lighting was established as the standard, although in some laboratories incandescent was required for technical reasons.

The approximate cost of one laboratory, including laboratory fittings, was $4·75 million; architectural fees and site utilities to within 5 ft of the building are excluded.

The relation between total staff and car-parking space is about 1·2 employees per space.

Chapter 9

What of the Future?

In the last decade many new materials have become available, and it is interesting to note their wide application in laboratories. Chemical-resistant vinyl, epoxy, polyester, and polyurethane finishes are used on benches, taps, walls, floors, ducts, etc., and plastic pipe has facilitated the reticulation of distilled water; tempered glass and rigid polyvinyl chloride pipes are used for laboratory effluent, and polypropylene traps of improved design and performance have replaced polythene. Although not new, items in which there have been considerable improvement include face bricks, exposed aggregate precast concrete panels, aluminium window frames, glass, prefabricated hot and cold rooms, stainless steel sinks, taps, and fume cupboard exhaust fans. As even more new materials come on the market and new building techniques develop, their use will result in still further improvements in design and performance.

In special laboratories and instrument rooms, it will be necessary for the air-conditioning to give improved distribution, individual room control, more efficient filtering, and finer tolerances of temperature and humidity. The high cost of some instruments will mean that they should be conveniently located for use by a maximum number of staff; in small laboratories this will not present a problem, but in larger projects this will best be met by increasing the area of each storey. The greater floor area in a double-width laboratory facilitates the sharing of equipment and communication between the scientists; also, internal laboratories simplify the problems of temperature variation and access to the more complex services and ducts.

In recent years the range of scientific equipment has increased enormously; one even reads of a 'robot chemist' which not only completely automates some elaborate chemical analyses, but also prints the results. Electron microscopes, mass spectrometers, and other complex instruments now are regarded as essential in most laboratories, and no doubt more and more scientific equipment will be used in the future. Whilst some will result in a reduction of staff and space, the overall effect is likely to be an increase in the number of instrument rooms and therefore more space per research worker.

As costs continue to rise, economics will demand the reduction of the on-site labour content of building construction and installation of laboratory fittings and services. High building costs will focus attention on design in relation to maximum flexibility and minimum maintenance, and the use of less expensive internal finishes will be encouraged. The cost of individually designed laboratory fittings will become so prohibitive that their use will be quite unjustified; the alternative is a complete acceptance of standard fittings manufactured by specialist firms.

Will research laboratories continue to function in much the same way as they do today? I think they will, and therefore I find it difficult to imagine any alteration to the basic concept of a laboratory building consisting of a series of rooms; the very nature of research work demands separate working laboratories, service laboratories, and laboratory offices.

Summarising, then, I leave you with this thought:

The field of laboratory planning offers plenty of scope for the designer of the future. The perfect laboratory has yet to be built!

Appendix 1

The Cost of Research

The subject of cost involved in maintaining a research establishment does not come within the scope of this book, but the following figures relating to CSIRO for the year ending June 1970 may be of interest:

1. Total expenditure was \$51·7 million—

Investigations	\$48·0 million
Grants to research associations, etc.	1·4 million
Buildings and acquisition of properties	2·3 million

The amounts expended in each of the past five years are shown below:

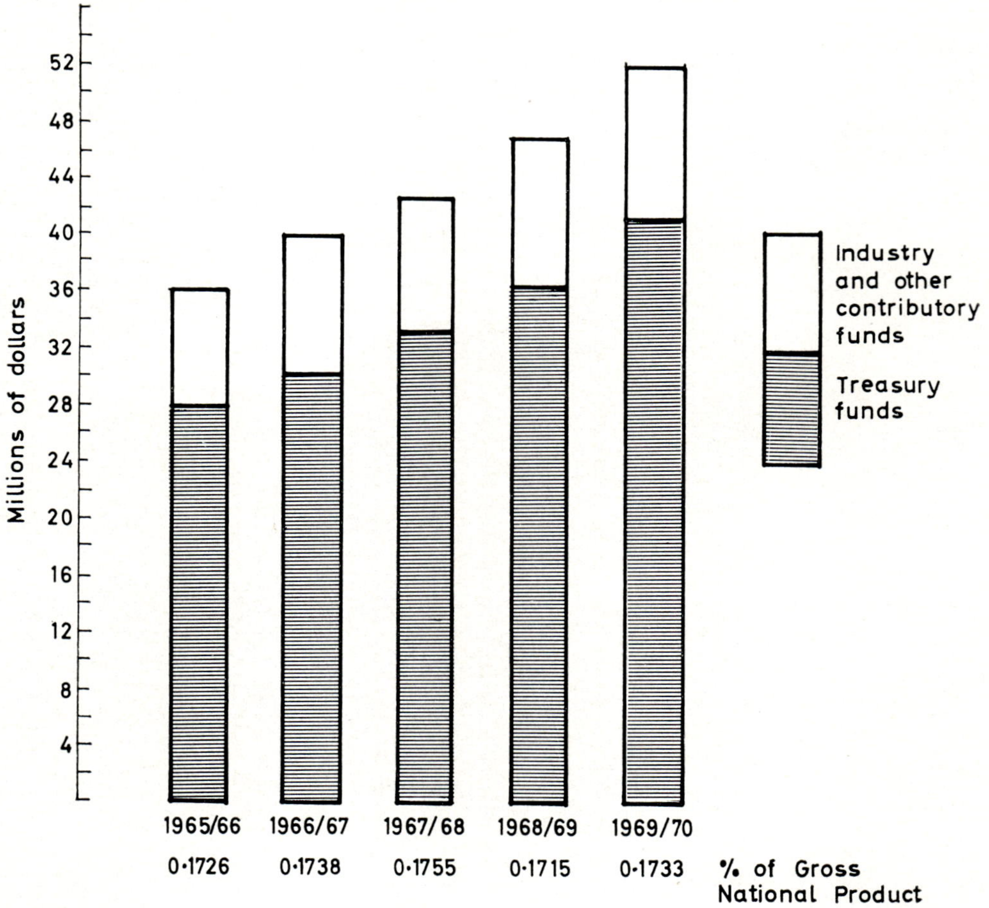

2. Approximately two-thirds of the $48·0 million spent on investigations relates to salaries; as shown below, the next most important item is maintenance—general running expenses:

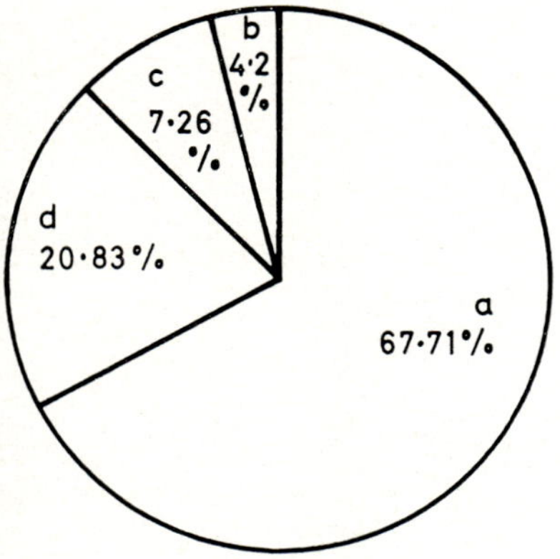

(a)	Salaries	$32·5 million
(b)	Travelling	2·0 million
(c)	Equipment	3·5 million
(d)	Maintenance	10·0 million

3. The total staff of 6400 was employed in the following categories:

Research scientists	1040
Experimental	750
Administrative	970
Other supporting staff	3120
Casual	520

4. One of the most difficult factors that must be taken into account in allocating resources is the continual rise in the cost of research. This is due in part to increases in salaries and costs of services; it is partly due also to the increasingly complex and expensive equipment required owing to the advances in science and techniques. Yet, if the Organisation is to remain effective in research, it must adopt these new developments as soon as they become available because they might save years of effort and at times may provide the only practicable means of tackling a problem. New techniques and new equipment also mean more skilled staff to operate and service them. The overall effect is that the annual cost of maintaining each research scientist is steadily rising; in 1962 it averaged $23 700, in 1965 it was $34 200, and in 1970 it was $46 100. A high proportion of the additional funds granted by the Government each year is absorbed by this type of inescapable increase in expenditure.

A Restricted Survey of Floor Loading

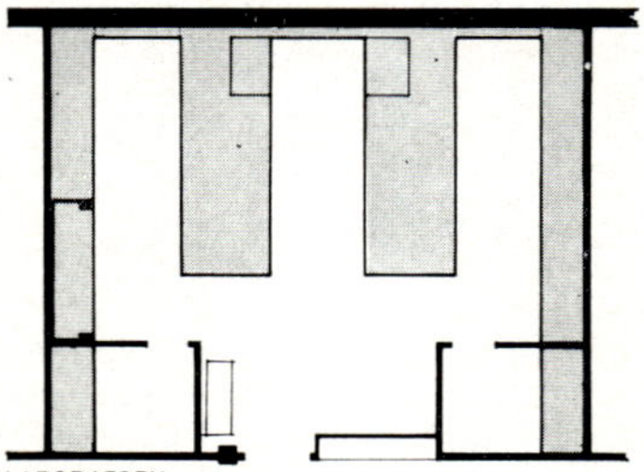

LABORATORY
Area – 700 ft^2
Weight of benches, etc. – 4,600 lb
Weight of equipment – 5,780 lb
Total weight – 10,380 lb = 15 lb / ft^2

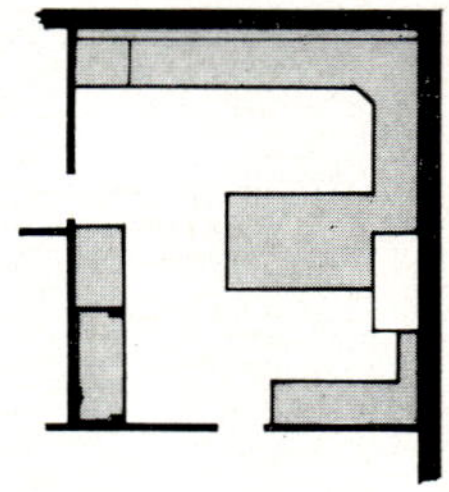

LABORATORY
Area – 330 ft^2
Weight of benches, etc. – 2,600 lb
Weight of equipment – 1,500 lb
Total weight – 4,100 lb = 13 lb/ft^2

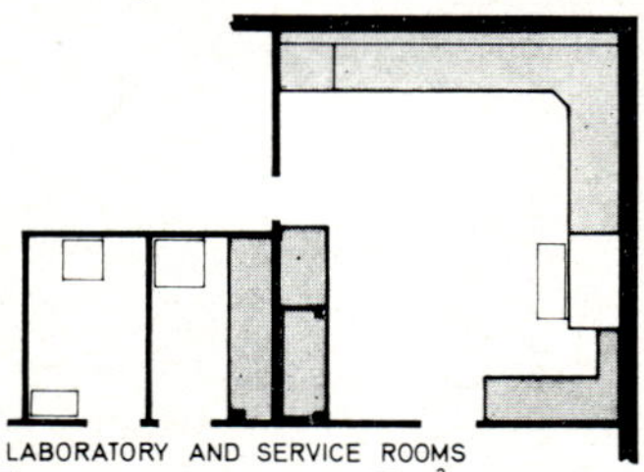

LABORATORY AND SERVICE ROOMS
Total area – 450 ft^2
Weight of benches, etc. – 2,465 lb
Weight of equipment – 2,355 lb
Total weight – 4,820 lb = 11 lb / ft^2

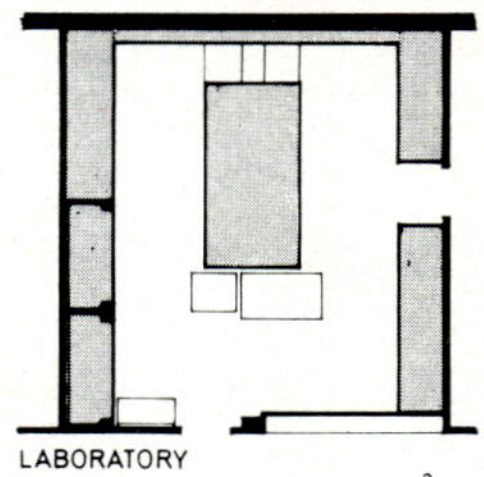

LABORATORY
Area – 380 ft^2
Weight of benches, etc – 3,130 lb
Weight of equipment – 3,570 lb
Total weight – 6,700 lb = 18 lb / ft^2

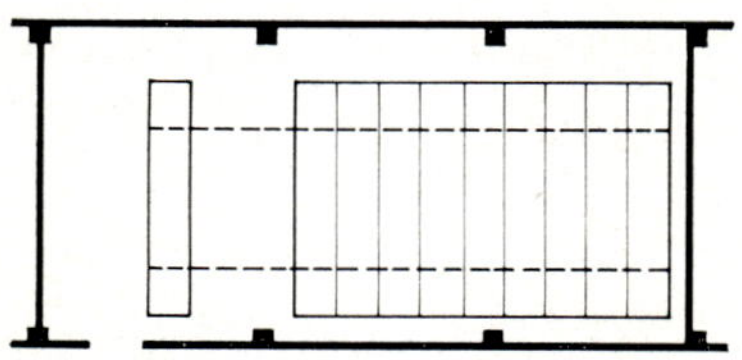

STACK ROOM
Area – 530 ft^2
Weight of 10 mobile units each 12ft 2in x 2ft 2in x 7ft 7½in high with 7 shelves loaded to capacity
Total weight 4,480 x 10 – 44,800 lb = 85 lb / ft^2
NOTE : Chemical Abstracts weighed 28·2 lb / lineal ft

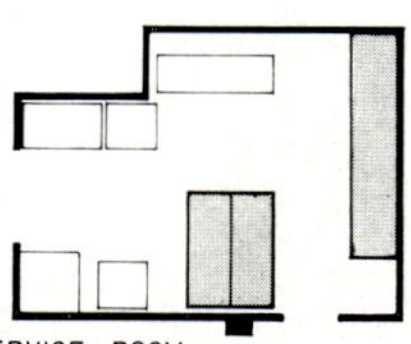

SERVICE ROOM
Area – 260 ft^2
Weight of benches, etc – 550 lb
Weight of equipment – 2,600 lb
Total weight – 3,150 lb = 12 lb / ft^2

Appendix 3

Aniline Bench Top Preparation

Solution 1	—	Copper sulphate	1 part	By weight
		Potassium chlorate	1 part	
		Water	8 parts	
Solution 2	—	A 5% solution of potassium bichromate		
Solution 3	—	Aniline	6 parts	By volume
		Hydrochloric acid	9 parts	
		Water	50 parts	

Using six coats in all, apply with a brush *in the following order*, allowing each coat to dry before commencing the next:

1, 2, 1, 2, 3, 3

Finish with raw linseed oil, well rubbed in; usually three coats are necessary.
One gallon of each of the above solutions will cover approximately 600 ft^2.
Bench units, walls, and floor should be protected to avoid splash marks. The back skirting should be fitted after the top is treated.

Index